Upper Elementary
Challenge Math

Edward Zaccaro

HICKORY GROVE PRESS

Enriching the Curriculum

About the Author

Ed lives outside of Dubuque, Iowa, with his wife Sara. He has been involved in various areas of education since graduating from Oberlin College in 1974. Ed holds a Masters degree in gifted education from the University of Northern Iowa and has presented at state and national conferences in the areas of mathematics and gifted education.

Library of Congress Control Number: 2014905961
ISBN 10: 0-9854725-2-9
ISBN 13: 978-0-9854725-2-8

Introduction

Meeting the needs of mathematically gifted children can be difficult. As teachers try to develop programs that provide appropriate challenges and also teach basic skills, decisions need to be made concerning acceleration, enrichment and differentiation. As these decisions are made, it is imperative that teachers also keep in mind that they must help students take intellectual risks; learn to think deeply and with insight; see the magic and wonders of mathematics and help students understand and appreciate mathematics and its place in the world. I have always found it helpful to consider the following points as decisions are made as to the most appropriate way to approach the needs of mathematically gifted children.

1) Challenge and frustration are a part of learning and life. They should both be viewed as a normal part of the learning process. While most mathematically gifted children enjoy challenging material, some children find the experience of challenge and frustration to be quite stressful because it is a foreign concept to them. Teachers of mathematically gifted children have the sometimes unpleasant task of helping these students understand that limiting their academics to an intellectual box where there is no struggle or frustration is not healthy and leads to a life that is not as fulfilling or as rewarding.

2) Math is often taught as all scales and no music. Children must have the opportunity to see the exciting and interesting parts of mathematics. The goal of many programs for mathematically gifted children is to move students through the curriculum as quickly as possible. This approach can lead to a loss of interest in the subject because it does not nurture a child's passion for mathematics. An alternative approach is to keep gifted children with their same age peers, but give them an opportunity to experience the parts of mathematics that are not only challenging, but also very interesting. When children first see the wonders of math and science, it is as if they stepped into a room that they didn't know existed.

3) It is important for children to be shown the fascinating connections between mathematics and the real world. Because mathematics instruction is often dominated by facts and calculation, children are rarely exposed to important concepts that connect math and science to the real world.

4) Children who are gifted in mathematics must learn to appreciate their gift. Can you imagine what it feels like for an athlete or musician to have hundreds of parents and classmates cheering for him or her? Add to that the newspaper articles, trophies, medals, and other awards. This kind of reinforcement pushes athletes and musicians to excel. It is unlikely that this kind of motivating environment will ever become routine for those students who excel in math and science. Because there are precious few opportunities for gifted children to be formally recognized and honored, it is important that teachers make students feel that their gifts are something to be treasured.

5) Parents and educators must understand that a child's interests and passions do not necessarily correspond with their areas of giftedness. Over the years of teaching mathematically gifted children, I developed an understanding of the importance of allowing and encouraging children to follow their passions, which may or may not be their area of giftedness.

6) Mathematically gifted children must be given material that truly challenges them and appropriately challenges them. Bright math students usually pick up concepts so quickly that they are left with very little to do intellectually while the rest of the class masters the new material. In addition, the consequences of not challenging elementary children can be serious because children who are bored tend to develop thinking skills and work habits that are less than ideal.

7) Highly able children must have the opportunity to work with children with similar abilities. The importance of having the opportunity to work with children of similar abilities cannot be overstated because the value of this kind of interaction is not limited to the intellectual growth that it can foster. The social and emotional development that can occur as a result of healthy disagreement, discussion, and debate can have a profound impact on mathematically gifted children. An additional benefit is a reduction in the social isolation that these children sometimes experience.

— *Ed Zaccaro* —

Books by Edward Zaccaro

- *Primary Grade Challenge Math*
- *Upper Elementary Challenge Math*
- *Challenge Math for the Elementary and Middle School Student*
- *Real World Algebra*
- *The Ten Things All Future Mathematicians and Scientists Must Know (But are Rarely Taught)*
- *Becoming a Problem Solving Genius*
- *25 Real Life Math Investigations That Will Astound Teachers and Students*
- *Scammed by Statistics: How We are Lied to, Cheated and Manipulated by Statistics*
- *Now You Know Volume 1*
- *Now You Know Volume 2*

Table of Contents

Chapter One
Astronomy, Light and Sound

I just found out a way to lose a lot of weight without eating less and exercising more!!

I think I know what you are going to say, but please tell me your idea.

Well, it isn't an easy way, but it is a way to get dramatic weight loss. I just read that because gravity on the moon is much less than on Earth, a 100 pound person would only weigh 17 pounds on the moon.

Remember that you are talking about weight. Even though you will weigh less on the moon, Venus or Pluto, you will still look the same. You need to understand the difference between weight and mass.

Your weight depends on the strength of the pull of gravity of the planet or moon you are on. Look at this brick. It weighs 10 pounds. If I sent it to the moon, it will look the same and it will still have all the "stuff" it is made of, but it will just weigh a lot less. Because the moon has weaker gravity than the Earth, the brick will only weigh about 1 ½ pounds.

Your weight will change if you go to other planets, but you will still contain the same amount of "stuff". The "stuff" that you are made of is called **matter.**

I see now that my strategy for losing weight would be futile. My weight will change on other planets, but the amount of matter I contain would stay the same.

It is very fun to study weights on other planets and stars. One of my favorite facts is that if I weighed myself on a neutron star, I would weigh billions of pounds, but if I weighed myself on Phobos, one of Mars' moons, I would only weigh ½₀ of a pound!

Find the missing weights for the following problems:

1) A cement block weighs 10 pounds on Earth and 1 ½ pounds on the moon. What would a cement block that weighs 5 pounds on Earth weigh on the moon?

2) If a 100 pound person weighed 20 pounds on an asteroid, how much would a 120 pound person weigh on the same asteroid?

3) A person who weighs 100 pounds on Earth would weigh 17 pounds on the moon. How many kilograms would a 250 pound person weigh on the moon? (Round to the nearest whole kilogram.)

You can change pounds into kilograms by multiplying by .45

For example: 50 pounds x .45 = 22.5 kilograms

I am 12 years old because the Earth has traveled 12 times around the sun since I was born. I wonder how old I am in Jupiter years.

Are you saying that you are a different age in Jupiter years?

To find out how old I am in Jupiter years, I need to know how many times Jupiter has traveled around the sun in 12 years. Look at this chart that tells how long it takes for each planet to travel once around the sun. (I have rounded each number.)

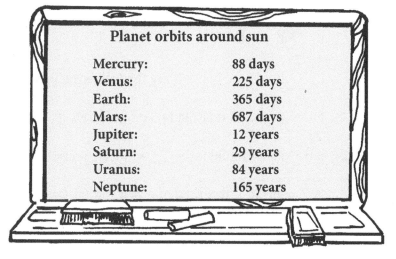

Planet orbits around sun

Mercury:	88 days
Venus:	225 days
Earth:	365 days
Mars:	687 days
Jupiter:	12 years
Saturn:	29 years
Uranus:	84 years
Neptune:	165 years

Now it is easy to see that you are only one in Jupiter years because it takes Jupiter 12 years to travel around the sun!

This is fun! I can find your age in Mercury years by finding how many days there are in 12 years. Then I divide by 88 days because it takes Mercury 88 days to go around the sun once.

12 x 365 = 4380 days

**4380 days ÷ 88 = 49.8
or about 50 Mercury years**

Wow! I am 50 in Mercury years. I should be able to get a driver's license on Mercury.

Try the following problems:

1) How old is a 58 year old in Saturn years?

2) An 8 year old child is _____ in Venus years. (Round to the nearest whole number.)

3) A 7 year old is less than one year old on Uranus. How many months old would a 7 year old be on Uranus?

Light and Sound

You can find out how far away a thunderstorm is by counting the time between seeing the lightning flash and hearing the thunder.

I know that lightning and thunder happen at the same time, but sound is very slow compared to lightning.

Sound takes approximately 5 seconds to travel one mile and we can say that the light from a lightning bolt travels instantly to our eyes because light is so fast that it can travel 7 times around the Earth in only one second.

So if I see lightning and hear thunder 15 seconds later, I know that the storm must be 3 miles away because it took sound 5 seconds to travel each mile.

 Sound travels ⅕ of a mile in one second while light travels 186,000 miles in one second. If we do the math, we find that while light is traveling 7 times around the Earth each second, it takes sound 125,000 seconds to travel once around the Earth. That is equal to almost a day and a half!

Try the following problems:

1) If thunder is heard 2½ seconds after lightning, how far away is the storm?

2) How far does light travel in one minute?

3) How far does sound travel in one minute?

Problem Set 1

Warmup: A 100 pound dog would weigh 17 pounds on the moon. A dog that weighs 50 pounds on Earth would weigh_____pounds on the moon.

Level 1: A person who weighs 100 pounds on Earth would weigh 236 pounds on Jupiter. How much would a person who weighs 25 pounds on Earth weigh on Jupiter?

Level 2: A person who weighs 100 pounds on Earth weighs 15 pounds on an asteroid. How much would a 120 pound person weigh on the asteroid?

Level 3: A 100 pound monkey would weigh 2800 pounds on the sun. How many kilograms would a 150 pound person weigh on the sun? (A pound is equal to .45 kilograms.)

Genius Level: A tortoise that weighs 100 pounds on Earth would weigh 38 pounds on Mercury. What would a tortoise that weighs 100 pounds on Mercury weigh on Earth? (Round to the nearest pound).

Problem Set 2

Warmup: How far does light travel in 20 seconds?

Level 1: The blink of an eye takes approximately ⅕ of a second. How far does light travel during the blink of an eye?

Level 2: How long would it take light to travel from the Earth to the moon? (The moon is 250,000 miles from the Earth.)

a) 1.3 seconds   b) 1.3 minutes   c) 1.3 hours   d) 1.3 days

Level 3: How many seconds are in a year?

Genius Level: How many miles does light travel in a year?

Problem Set 3

Warmup: The moon's diameter is approximately 2000 miles and the Earth's diameter is 8000 miles. The moon's diameter is what fraction of the Earth's diameter?

Level 1: The Earth's diameter is 8000 miles and Jupiter's diameter is 89,000 miles. How many times larger is Jupiter's diameter than the Earth's diameter? (Round to the nearest whole number.)

Level 2: The length of a day on Mars is 24 hours 37 minutes and 22 seconds. The length of a day on Earth is 23 hours, 56 minutes and 4 seconds. How much longer is a Mars day than an Earth day?

Level 3: The Earth's circumference at the equator is approximately 25,000 miles. At what speed (in mph) does the Earth rotate if it turns once in 24 hours?

Genius Level: Jupiter's circumference is approximately 280,000 miles at the equator. The length of its day is 9 hours, 50 minutes and 28 seconds. At what speed does Jupiter rotate in miles per second? (Round to the nearest whole number.)

Problem Set 4

Warmup: You see lightning and then hear thunder 5 seconds later. How far away is the storm? (Remember that lightning and thunder happen at the same time.)

Level 1: You are standing across a river from a friend. Through binoculars, you see your friend's lips move but don't hear anything for 2½ seconds. How far away is your friend?

Level 2: You are standing some distance from a large rock wall. When you shout, it takes 10 seconds for you to hear an echo. How far away is the rock wall?

Level 3: How far away is an electrical storm if you see lightning and then hear thunder 8 seconds later?

Genius Level: During an electrical storm, you see lightning and then hear thunder 15 seconds later. Ten minutes later, the lightning and thunder occur at the same time. How fast is the storm moving?

Problem Set 5

Warmup: How far does light travel in ½ second?

Level 1: How far does light travel in one minute?

Level 2: How many times can light travel around the equator of the Earth in one second?

    a) .5 times   b) about 2 times   c) 7.5 times   d) 750 times

Level 3: The sun is 93,000,000 miles from Earth. How long does it take light to travel from the sun to the Earth? (Express your answer in minutes and seconds.)

Genius Level: Light travels 186,000 miles per second and sound travels ⅕ of a mile in one second. How many times faster is light compared to sound?

  a) 10 times faster   b) 100 times faster   c) 1,000 times faster   d) 1,000,000 times faster

Problem Set 6

Warmup: A person who is 36 years old is how old in Jupiter years?

Level 1: A 41 year old is approximately how old in Neptune years?

      a)1 year old   b) 4 years old   c) ¼ year old   d) 1 month old

Level 2: The voting age in the United States is 18 years old. How old is that in Venus years? (Round to the nearest whole number.)

Level 3: The President of the United States must be at least 35 years old. How old is that in Mars years? (Round to the nearest tenth.)

Genius Level: An Earth day is approximately 24 hours because the Earth rotates once every 24 hours. Jupiter rotates once in approximately 10 hours. How many "Jupiter days" are in one "Jupiter year"?

Astronomy
Level 1

1) How far does sound travel in one minute?

2) A person who weighs 34 pounds on the moon would weigh _____ on Earth.

      a) 100 pounds   b) 125 pounds

      c) 175 pounds   d) 200 pounds

3) The moon's orbit around the Earth is approximately 1,500,000 miles. The Earth's orbit around the sun is approximately 585,000,000 miles. How many times longer is the distance of the Earth's orbit around the sun compared to the moon's orbit around the Earth?

4) Belinda is standing on the top of a hill. She looks across a valley at another hill and sees lightning strike a large tree. 15 seconds later, she hears thunder. How far away was the lightning strike?

5) One astronomical unit is the distance from the Earth to the sun, which is approximately 93,000,000 miles. Three astronomical units would be 3 x 93,000,000 = 279,000,000 miles. How many astronomical units is Jupiter from the sun if it is 484,000,000 miles from the sun? (Round to the nearest tenth.)

Astronomy

Level 2

1) If you hear thunder 17½ seconds after a lightning strike, how far away is the electrical storm?

2) The speed of sound in water is how much faster than the speed of sound in air? *(Research needed)*

   a) About half the speed   b) Same speed   c) Twice as fast   d) 4 times faster

3) What fraction of a second does it take light to travel around the Earth's equator once?

   a) ½ second   b) ¼ second   c) ⅐ second   d) ¹⁄₁₀ second

4) The Earth travels 585,000,000 miles around the sun in one year and there are approximately 8760 hours in a year. What is the Earth's average speed in miles per hour as it travels around the sun? (Round to the nearest whole number.)

5) A light-year is the distance light travels in a year. If a sloth moves at an average speed of 7200 feet per day, how many miles are in a sloth-year?

   a) 50 miles   b) 100 miles   c) 200 miles   d) 500 miles

Astronomy
Level 3

1) It takes light from the sun between 8 and 9 minutes to reach Earth. How long does it take light from the sun to reach Pluto? *(Research needed)*

   a) 5 ½ minutes   b) 5 ½ hours   c) 5 ½ days   d) 5 ½ years

2) If you weigh 100 pounds on Earth, what would your weight be on a typical neutron star? *(Research needed)*

   a) ¹⁄₁₀₀ pound   b) 100,00 pounds   c) 1-2 million pounds   d) 1 trillion pounds

3) The Earth's distance to the moon is what fraction of the Earth's distance from the sun? *(Research needed)*

   a) ¹⁄₄₀₀   b) ¹⁄₁₀₀   c) ¹⁄₅₀   d) ¹⁄₁₀

4) The Earth's escape velocity is the speed an object must travel in order to escape the Earth's gravity. The escape velocity of Earth is approximately 7 miles per second. What is the Earth's escape velocity in miles per hour?

5) After seeing lightning, thunder is heard 15 seconds later. 10 minutes later, thunder is heard 10 seconds after the lightning strike. How fast is the storm moving?

Astronomy
Genius Level

1) The Earth's escape velocity is how many times faster than the speed of sound?

    a) Twice as fast   b) 15 times as fast   c) 35 times as fast   d) Both are about the same speed

2) During an electrical storm, you see lightning and then hear thunder 30 seconds later. 12 minutes later, the thunder and lightning occur at the same time. How fast is the storm moving?

3) Sound and light start a race at the bottom of the Empire State Building. The race is to the moon and back. Where will sound be after one second? Where will light be after one second? For each question, choose from the 6 possible answers below:

  a) Between 100 and 200 miles from Earth   d) Finished the race

  b) Almost to the moon   e) Reached the moon and heading back

  c) Almost to the top of the Empire State Building   f) About 6 miles from Earth

4) Jets break the sound barrier when they travel faster than the speed of sound. What is the speed a jet must travel to break the sound barrier?

    a) 160 mph   b) 360 mph

    c) 760 mph   d) 1760 mph

5) The moon travels at 2288 miles per hour around the Earth. It travels approximately 1,500,000 miles each time it completes a journey around the Earth. How long does it take the moon to travel around the Earth?

    a) 14 days   b) 27.3 days

    c) 30.1 days   d) 31.3 days

Chapter Two
Problem Solving

1) There is a jar with marbles inside. If Anna can guess the number of marbles in the jar, she will win free movie passes for a year. Here are her clues:

a) ½ of the marbles are red
b) ¼ of the marbles are green
c) ⅛ of the marbles are blue
d) There are 50 remaining marbles that are yellow

2) If hotel rooms 200 through 650 are cleaned, how many rooms were cleaned?

3) It takes Jared 2 hours to paint a fence. Ian and Matthew each can paint the same fence in 4 hours. If they work together, how long will it take to paint the fence?

These problems are impossible! My brain just can't figure out what to do to solve them!

The brain needs some extra help when trying to solve problems such as these. Here is a box of problem-solving strategies that will help you.

Draw a Picture
2-10 Method
Think 1

Thanks! I hope they help because I really need it!

Draw a Picture

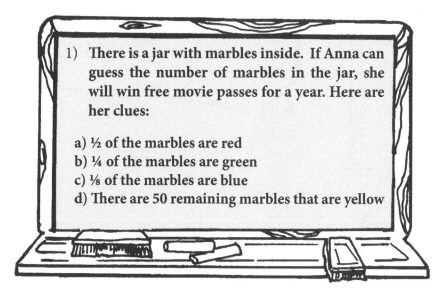

1) There is a jar with marbles inside. If Anna can guess the number of marbles in the jar, she will win free movie passes for a year. Here are her clues:

a) ½ of the marbles are red
b) ¼ of the marbles are green
c) ⅛ of the marbles are blue
d) There are 50 remaining marbles that are yellow

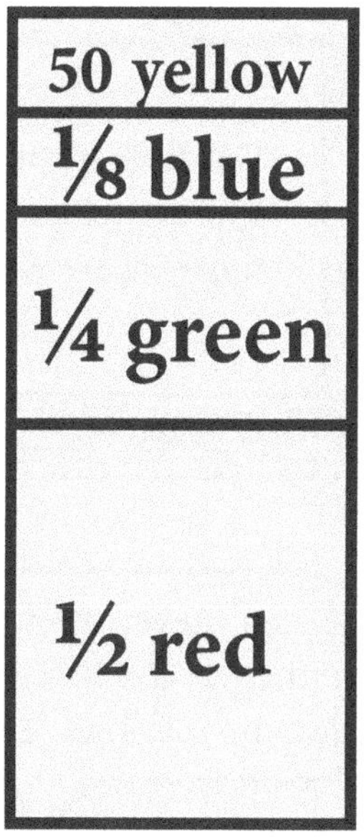

| 50 yellow |
| ⅛ blue |
| ¼ green |
| ½ red |

Step 1: Draw a rectangle for the jar

Step 2: Divide the rectangle and write in what you know

Step 3: It is now clear that the 50 yellow marbles are the remaining ⅛ of the jar

Step 4: If ⅛ of the jar = 50 marbles, then the jar must contain 8 x 50 marbles = 400 marbles

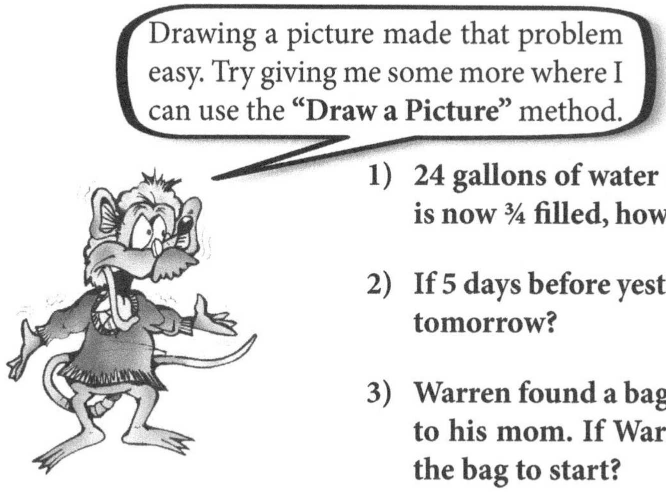

Drawing a picture made that problem easy. Try giving me some more where I can use the **"Draw a Picture"** method.

1) 24 gallons of water are poured into an empty tank. If the tank is now ¾ filled, how many gallons of water does the tank hold?

2) If 5 days before yesterday is a Saturday, what day of the week is tomorrow?

3) Warren found a bag of money. He gave ½ to his brother and ⅙ to his mom. If Warren has $80 left, how much money was in the bag to start?

2-10 Method

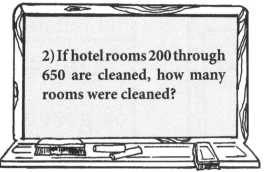

2) If hotel rooms 200 through 650 are cleaned, how many rooms were cleaned?

I can never remember what the rule is for this type of problem. I know the answer isn't just 650 - 200 = 450 rooms.

I can never remember the rule either. Is it 450 + 2 = 452 rooms? Maybe it is 450 - 1= 449 rooms.

In this situation, it is easy to find the rule by plugging in the numbers 2 and 10. Now the problem becomes:

If hotel rooms 2 through 10 are cleaned, how many rooms were cleaned?

That problem solving technique will make this easy. I couldn't count the rooms before because there were too many. Now it is a small number of rooms:
The rooms that were cleaned are 2, 3, 4, 5, 6, 7, 8, 9, and 10.

There were 9 rooms cleaned. 10 - 2 = 8 so I know the rule is subtract and then add one.
650 - 200 = 450 + 1 = 451

 You can also use the **2 - 10 method** when there are fractions, decimals or even variables in the problem. Using a 2 and a 10 makes it easier for the brain to think. The problem below is very, very confusing until you use the **2 - 10 method**.

I remember what a variable is! It is a letter that stands for a number you don't know.

I think of variables as mystery numbers.

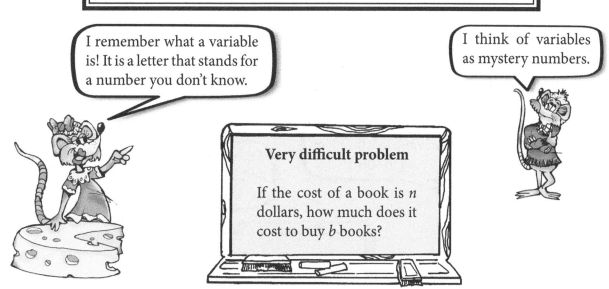

Very difficult problem

If the cost of a book is *n* dollars, how much does it cost to buy *b* books?

2 - 10 Method:

If the cost of a book is **2** dollars, how much does it cost to buy **10** books?

 When using the 2 - 10 method, substitute a 10 for the larger number and a 2 for the smaller number.

When a **2** and a **10** are plugged in, it is easy to see that the answer is 2 x 10 = $20.

Now we know how to solve the problem— We need to multiply.

Real problem: *n* x *b* is the answer

1) If 2.25 inches on a map are equal to 18 miles, what is one inch equal to?

2) How many ¾ pound pieces can be cut from a giant 1200 pound chocolate Easter Bunny?

3) There are *n* ostriches and *t* horses on a farm. How many legs are there?

Think 1

3) It takes Jared 2 hours to paint a fence. Ian and Matthew each can paint the same fence in 4 hours. If they work together, how long will it take to paint the fence?

I'm thinking of 1, but it isn't helping.

If you use the **Think 1** method of problem solving, this problem becomes very easy to solve.

I think you are supposed to think of 1 hour. That really helps. Look at how I solve the problem!

In 1 hour Jared: ½ the fence

In 1 hour Matthew: ¼ the fence

In 1 hour Ian: ¼ the fence

Now I'll draw a picture. I can see that they will paint the whole fence in 1 hour.

| Jared ½ | Matthew ¼ | Ian ¼ |

½ + ¼ + ¼ = 1

1) It took 2 people 3 hours to build a snow fort. If one more friend helps, how long would it take 3 people to build the snow fort? (Hint: How many hours would it take one person to build the snow fort?)

2) Hose A will fill a large tank in 2 hours. Hose B will fill the tank in 4 hours and Hose C will fill the tank in 4 hours. If all 3 hoses are turned on, how long will it take to fill the tank?

3) Jay has 3 cats that need to be fed while he is on a 2-week vacation. If one bag of food will feed 2 cats for 7 days, how many bags of food does Jay need for his 14 day vacation?

Problem Set 1
(Draw a Picture)

Warmup: If tomorrow is Sunday, what day of the week is yesterday?

Level 1: Four gallons of gas are put into a gas tank that was ½ full. Now it is ¾ full. How many gallons of gas does the entire tank hold?

Level 2: Three days after yesterday is Friday. What day of the week is today?

Level 3: Laura found a bag of money. She gave ½ to charity, ¼ to her brother, and ⅛ to a friend. She now has $25 left. How much money was originally in the bag?

Genius Level: A gas tank is ⅜ full. When 3 gallons are added, the tank is 9/16 full. How many gallons does a full tank hold?

Problem Set 2
(2-10 Method)

Warmup: If the price of chicken is $3.25 per pound, how much does 8 pounds of chicken cost?

Level 1: How many ¼ pound pieces can be cut from a giant 3 pound chocolate Easter Bunny?

Level 2: If there are n pigs on a farm and p ducks, how many total legs are there?

Level 3: If hotel rooms 100 through 795 were cleaned, how many rooms were cleaned?

Genius Level: If it takes m minutes to read p pages, how long does it take to read one page?

Problem Set 3
(2-10 Method)

Warmup: If apples cost $1.50 per pound, how much would 1.5 pounds of apples cost?

Level 1: If there are 12 pizzas and 72 people, what fraction of a pizza would each person receive?

Level 2: If .75 inches on a map are equal to 12 miles, how many miles is one inch equal to?

Level 3: Juan can read ¾ of a page in one minute. How long will it take him to read 72 pages?

Genius Level: If the cost of each frog is n dollars, how much does it cost to buy f frogs?

Problem Set 4
(Think 1)

Warmup: If it takes 4 hours to paint a fence, what fraction of the fence is painted in one hour?

Level 1: If it takes Trisha 12 hours to paint a fence, what fraction of the fence is painted in one hour?

Level 2: It takes Daniel 2 hours to paint a fence and Alicia 2 hours to paint the same fence. How long will it take to paint the fence if they work together?

Level 3: It takes Luke 2 hours to paint a fence and Laura 4 hours to paint a fence. It takes Jon 4 hours to paint the fence. If all three work together, how much of the fence will they paint in one hour?

Genius Level: Hose *A* takes 4 hours to fill the town swimming pool, while hose *B* takes 2 hours to fill the town swimming pool. If both hoses are turned on, how long will it take them to fill the pool?

Problem Set 5
(Think 1)

Warmup: If it took 3 people 10 hours to plant a large garden, how long would it take one person?

Level 1: If it takes 2 people three hours to pick 20 rows of corn, how long would it take 3 people?

Level 2: If it takes three snowplows 12 hours to plow a highway, how long would it take 4 snowplows?

Level 3: If it takes 3 people 8 hours to wash 15 cars, how long would it take 2 people to wash 7½ cars?

Genius Level: If 3 equally sized hoses fill a pool in 3 hours and 20 minutes, how long would it take 2 hoses to fill the pool?

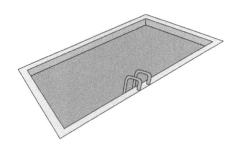

Problem Solving
Level 1

1) When 3 people mow a football field, it takes 8 hours. How long would it take 4 people to mow the football field?

2) Three gallons of gas are put into a gas tank that was ⅗ full. Now it is ⅘ full. How many gallons does the entire gas tank hold?

3) If yesterday was Sunday, what day of the week is tomorrow?

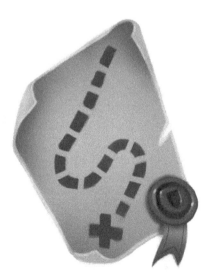

4) If .6 inches on a map equals 15 miles, how many miles is one inch equal to?

5) If 12 pizzas were bought for 96 people, how much pizza would each person receive?

Problem Solving
Level 2

1) If it takes 7 people 10 hours to mow a football field, how long would it take 5 people?

2) 2¼ gallons of gas are put into a gas tank that was half full. Now the tank is ⅝ full. How many gallons of gas does the entire tank hold?

3) The day after tomorrow is Saturday. What day of the week was yesterday?

4) There are *p* pigs and *y* chickens in a truck. How many legs are in the truck?

5) The cost of each baseball is 8 dollars. What is the cost of *n* baseballs?

Problem Solving

Level 3

1) When 3 people mow a football field, it takes 6 hours and 40 minutes. How long would it take 8 people to mow the football field?

2) 2⅝ gallons of gas are put into a gas tank that was ⅜ full. The tank is now ⁹⁄₁₆ full. How many gallons of gas does the entire tank hold?

3) If Jacob reads 1.4 pages per minute, how long will it take him to read 126 pages?

4) The sound from a thunderstorm travels approximately ⅕ of a mile in one second. How far will sound travel in 22.5 seconds?

5) Lauren wants to fence in a square garden to keep out rabbits. The garden will be 25 feet on each side. If Lauren places posts every 5 feet, how many posts will she need?

Problem Solving
Genius Level

1) It takes Abe 6 hours to shovel his driveway. It takes Lana 3 hours to shovel the same driveway. If they work together, how long will it take them to shovel the driveway?

2) It takes 30 people 8 days to build a brick wall. If 10 people are added to the work crew, how many days will it take for the wall to be completed?

3) If *n* airplanes can be built in *y* days, how long does it take to build 2 airplanes?

4) A tank is ¾ full. When 6 gallons are added, the tank is 99% full. How many gallons does the full tank hold?

5) Lynn can type *p* pages in *m* minutes. How long does it take Lynn to type one page?

Chapter Three
Sequences, Place Value, and Bases

One of my hobbies is guessing a missing number in a sequence.

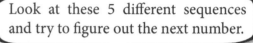

Look at these 5 different sequences and try to figure out the next number.

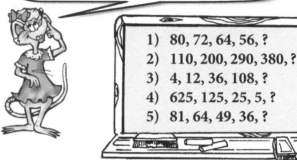

1)   80, 72, 64, 56, ?
2)   110, 200, 290, 380, ?
3)   4, 12, 36, 108, ?
4)   625, 125, 25, 5, ?
5)   81, 64, 49, 36, ?

I can find the answers to the first 4 problems, but the last one has me flummoxed.

1) **48**   Subtract 8 each time: 56 - 8 = 48

2) **470**   Add 90 each time   380 + 90 = 470

3) **324**   Multiply by 3 each time   108 x 3 = 324

4) **1** Divide by 5 each time   5 ÷ 5 = 1

5) 81, 64, 49, 36, ?

I am impressed that you used the word flummoxed instead of just saying you were confused or perplexed!

Some sequences can't be figured out just by thinking about adding, subtracting, multiplying, or dividing. Sometimes you have to be very clever and carefully look at the numbers.

I think I see now! 81, 64, 49 and 36 are all special numbers. The next number must be 25! Try the 3 problems shown below.

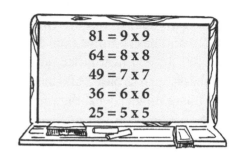

81 = 9 x 9
64 = 8 x 8
49 = 7 x 7
36 = 6 x 6
25 = 5 x 5

1) 15, 9, 3, -3, ?

2) ¹⁄₁₆, ⅛, ¼, ½, ?

3) 125, 25, 5, 1, ?

Place Value

The place value of numbers is something every student must understand. We'll pull back the curtain and look at the number 72,846 and see what each of the numbers mean.

6 is in the one's column   6 x 1 = 6
4 is in the 10's column   4 x 10 = 40
8 is in the 100's column   8 x 100 = 800
2 is in the 1000's column   2 x 1000 = 2000
7 is in the 10,000's column   7 x 10,000 = 70,000

So the number 72,846 is another way of saying:
70,000 + 2000 + 800 + 40 + 6

This is the expanded form of a number.

1) In the number 642,379, what number is in the 10,000 column? What number is in the 1000 column?

2) 690 written in expanded form is 600 + 90. Write 99,999 in expanded form.

3) In the number 9,876,543,210 the 6 is in the million's column. What column is the 8 in? What column is the 9 in?

I know our number system is base 10 and has columns such as 1000, 100, 10, 1. I heard there are other bases such as base 2 and base 7 but I have no idea what they are.

Bases that are different from base 10 are very easy to understand once you know they have different columns.

In base 10, we start with the 1's column and then we find other columns by multiplying by 10. The largest digit is one less than 10, which is 9.

1's column
1 x 10 = 10's column
10 x 10 = 100's column
10 x 100 = 1000's column
10 x 1000 = 10,000's column

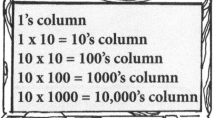

In base 7, we start with the 1's column and then we find other columns by multiplying by 7. The largest digit is one less than 7, which is 6.

1's column
1 x 7 = 7's column
7 x 7 = 49's column
7 x 49 = 343's column
7 x 343 = 2401's column

Look at what the numbers mean in the base 7 number 341.

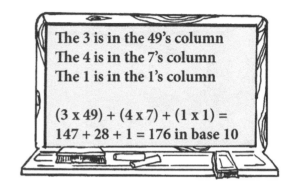

The 3 is in the 49's column
The 4 is in the 7's column
The 1 is in the 1's column

(3 x 49) + (4 x 7) + (1 x 1) =
147 + 28 + 1 = 176 in base 10

I get it now! I bet the columns for base 6 would be 1, 6, 36, 216 and the largest digit is a 5. The base 6 number 432 would be:
(4 x 36) + (3 x 6) + (2 x 1) =
144 + 18 + 2 = 164 in base 10
Bases aren't as hard as I thought they would be.

1) Base 7 number 24,532   What does the 4 stand for?

2) Base 7 number 555   Write in expanded form.

3) Base 7 number 60,000   If you had this many marbles in a collection, how many marbles would you have in base 10?

To change numbers from different bases into base 10, just use expanded form. Look at how I change the following base 7 numbers into base 10.

Base 7 columns: 2401, 343, 49, 7, 1

Base 7: 543   This means (5 x 49) + (4 x 7) + (3 x 1) = 276

Base 7: 12,345   This means:
(1 x 2401) + (2 x 343) + (3 x 49) + (4 x 7) + (5 x 1) = 3267

I see! So if an alien from the planet Septon said he had 11 children, I would have to convert the base 7 number into base 10.

Base 7 columns: 49, 7, 1

Base 7: 11 This means:
1 group of 7 + 1 group of 1 = 8

He has 8 children

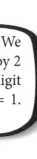

Let's do the same type of thinking for base 2. We start with the 1's column and then multiply by 2 to find each additional column. The largest digit possible in base 2 of course would be 2-1 = 1. The only digits used in base 2 are 1 and 0.

Base 2 columns: **64, 32, 16, 8, 4, 2, 1**

Look at how I can use expanded form to show what base 2 numbers are. This is exciting!

Base 2 columns: 64, 32, 16, 8, 4, 2, 1

Base 2 number: 111   **This means:**
1 group of 4 + 1 group of 2 + 1 group of 1 = 7

Base 2 number: 10,010   **This means:**
(1 x 16) + (0 x 8) + (0 x 4) + (1 x 2) + (0 x 1) = 18

Base 2 number: 100   **This means: (1 x 4) + (0 x 2) + (0 x 1) = 4**

1) In the base 2 number 10,000, what does the 1 stand for?

2) Write the base 2 number 10,101 in expanded form.

3) If Marcus said that he had $1,000,000 in base 2, how much money does he have in base 10?

Look at the place values for other bases.

Base 10: 10,000, 1000, 100, 10, 1
Base 9: 6561, 729, 81, 9, 1
Base 8: 4096, 512, 64, 8, 1
Base 7: 2401, 343, 49, 7, 1
Base 6: 1296, 216, 36, 6, 1
Base 5: 625, 125, 25, 5, 1
Base 4: 256, 64, 16, 4, 1
Base 3: 81, 27, 9, 3, 1
Base 2: 16, 8, 4, 2, 1

Problem Set 1

Warmup: 60, 45, 30, 15, 0, ?   What is the next number?

Level 1: 395, 300, 205, 110, 15, ?   What is the next number?

Level 2: 90, 45, 22½, 11¼, ?   What is the next number?

Level 3: 100, 10, 1, ¹⁄₁₀, ?   What is the next number?

Genius Level: 625, 576, 529, 484, ?

Problem Set 2

Warmup: 1½, 3, 4½, 6, 7½, ?   What is the next number?

Level 1: 100, 50, 25, 12.5, ?   What is the next number?

Level 2: ½, .25, ⅛, .0625, ?   What is the next number?

Level 3:   1   4   9   16 ....?   What is the 100th number?
            1st 2nd 3rd 4th   100th

Genius Level: 1, 8 27, 64, ?   What is the next number?

Problem Set 3

Warmup: 11, 7, 3, ?   What is the next number?

Level 1: 800, 200, 50, ?   What is the next number?

Level 2: One trillion, one billion, one million, ?   What is the next number?

Level 3: One trillion, one billion, one million, ?, ?, ?   What are the next three numbers?

Genius Level: 2, 5, 10, 17, 26 The next number is 37. What is the 40th number in the sequence?

Problem Set 4

Warmup: 7432 is 2 groups of one; 3 groups of ten; 4 groups of 100 and 7 groups of _____

Level 1: What does the 6 stand for in the number 761,852 ?

Level 2: Base 7 columns: 1, 7, 49, ?   What is the next column?

Level 3:
Base 10 columns: 1, 10, 100, 1000
Base 5 columns: 1, 5, 25, 125
Base 12 columns: ?, ?, ?, ?

Genius Level: 1, 10, 11, 100, 101, 110, 111, ?   What is the next number?
(Hint: Numbers are not base 10.)

Problem Set 5

Warmup: Base 5 columns: 1, 5, 25, 125   What are the first four base 6 columns?

Level 1: 1, 1⅞, 2¾, 3⅝, ?   What is the next number?

Level 2: 900, 90, 9, ?   What is the next number?

Level 3: 4, 2, 1, .5, .25, ?, ?, ?   What are the next three numbers?

Genius Level: 0, 1, 1, 2, 3, 5, 8, 13, 21, 34, ?   What is the next number?

Sequences, Place Value and Bases
Level 1

1) $10, $5, $2.50, ?　　　What is the next number?

2) ¼, ¾, 1¼, 1¾ ?　　　What is the next number?

3) In the number 65,431, the number 3 means there are 3 groups of 10 or 30. What does the 5 stand for?

4) .25, .75, 1.25, 1.75, ?　　　What is the next number?

5) 1, 4, 9, 16, ?　　　What is the next number?

Level 2

1) In the number 843,529 the 2 stands for 2 groups of 10 or 20. What does the 8 stand for?

2) 66, 55, 44, 33, 22, 11, ?　　　What is the next number?

3) Put in the missing columns for base 2:

Base 10 columns are	100,000	10,000	1000	100	10	1
Base 2 columns are	?	?	?	4	2	1

4) 21, 14, 7, 0, ?　　What is the next number?

5) 1, 4, 9, 16, ?　　　What is the 10th number?

Sequences, Place Value and Bases
Level 3

1) 64, 16, 4, 1, ? What is the next number?

2) 1, 3, 7, 15, 31, ? What is the next number?

3) 100, 81, 64 ? What is the next number?

4) On December 1st, a frog is normal size.
    On December 2nd, the frog is 4 times its normal size.
    On December 3rd, the frog is 9 times its normal size.
    On December 4th, the frog is 16 times its normal size.

    On what day in December will the frog be 400 times its normal size?

5) On December 1st, Isaac is given $2. On December 2nd he is given $4. On December 3rd, Isaac is given $8; $16 on the 4th; $32 on the 5th and this continues. On what day in December will Isaac be given $1024?

Genius Level

1) 1, 2, 6, 24, 120, 720, ?   What is the next number?

2) Fill in the missing base 9 columns:

Base 10 columns:	1000	100	10	1	.	$\frac{1}{10}$	$\frac{1}{100}$
Base 9 columns:	729	?	9	1	.	?	?

3) $\frac{1}{256}$, $\frac{1}{64}$, $\frac{1}{16}$, ?, ?, ?, ?   What are the next 4 numbers?

4) Trillion dollars, Billion dollars, Million dollars, Thousand dollars, ?
    What is the next amount of money?

5) 2, 3, 5, 7, 11, 13, ? What is the next number?

Chapter Four
Decimals

We just learned about place value. The number 7632 means 7000 + 600 + 30 + 2. I also see numbers with a decimal point. What would the number 641.27 mean?

Numbers after the decimal point also have place values. Look at the place values for base 10.

Base 10 place values

1000 100 10 1 • $\frac{1}{10}$ $\frac{1}{100}$

That helps a lot! That means the number 641.23 in expanded form is equal to:

$$600 + 40 + 1 + \frac{2}{10} + \frac{3}{100}$$

If you pretend 641.23 is money, it is even easier to understand.

$100		
$100	$10	
$100	$10	
$100	$10	$1
$100	$10	
$100	$10	
$100		

1) Write the number 38.34 in expanded form.

2) Write the number 512.111 in expanded form.

3) There are also decimals in different bases. What are the first two place values after the decimal point in base 5?

    125   25   5   1   .   ?   ?

4) Write the base 4 number 123.31 in expanded form.

5) Fill in the missing base 2 place value columns.
    32 ? ? ? 2 1 . ? ¼ ? ?

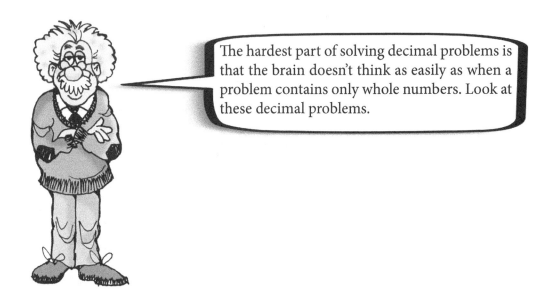

The hardest part of solving decimal problems is that the brain doesn't think as easily as when a problem contains only whole numbers. Look at these decimal problems.

1) The price of salmon at a fish market is $9.25 per pound. What is the cost of 3.2 pounds?

2) Adrian's pet cat weighs 10.35 pounds and his pet fish Jamison weighs .2 pounds. How much more does Adrian's cat weigh than his pet fish?

3) If Maria paid $81.60 for 6.4 pounds of fish, what did the fish cost per pound?

4) Adrian stepped on a scale holding his pet cat. If Adrian weighs 62.65 pounds and the cat weighs 10.35 pounds, what is their total weight on the scale?

I can easily tell when a decimal problem is addition or subtraction, but I need to use the 2-10 method when it is a multiplication or a division problem.

Me too! Look at the first problem after I changed it using the 2-10 method.

1) The price of salmon at a fish market is $2 per pound. What is the cost of 10 pounds?

Now it is easy to see that I need to multiply. I put the real numbers back in the problem and multiply.

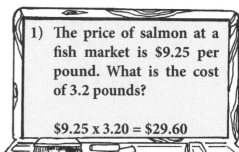

1) The price of salmon at a fish market is $9.25 per pound. What is the cost of 3.2 pounds?

$9.25 x 3.20 = $29.60

2) Adrian's pet cat weighs 10.35 pounds and his pet fish Jamison weighs .2 pounds. How much more does Adrian's cat weigh than his pet fish?

```
  10.35
-   .20
  10.15
```

Adrian's cat weighs 10.15 pounds more than his fish.

Remember to keep decimal points lined up when you add or subtract decimals. Look at problems #2 and #4.

4) Adrian stepped on a scale holding his pet cat. If Adrian weighs 62.65 pounds and the cat weighs 10.35 pounds, what is their total weight on the scale?

```
  62.65
+ 10.35
  73.00
```

Together they weigh 73 pounds.

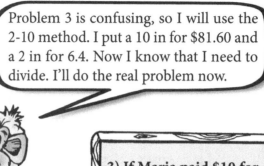

Problem 3 is confusing, so I will use the 2-10 method. I put a 10 in for $81.60 and a 2 in for 6.4. Now I know that I need to divide. I'll do the real problem now.

3) If Maria paid $10 for 2 pounds of fish, what did the fish cost per pound?

When you use the 2-10 method, remember to substitute a 10 for the larger number and a 2 for the smaller number.

3) If Maria paid $81.60 for 6.4 pounds of fish, what did the fish cost per pound?

$81.60 ÷ 6.4 = $12.75
The fish cost $12.75 per pound

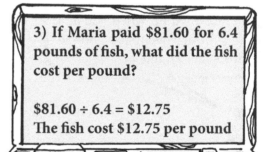

Problem Set 1

Warmup: A store charges $2.50 for two feet of rope. What is the cost per foot?

Level 1: A 10 foot rope will be cut into pieces 1.25 feet long. How many pieces will there be?

Level 2: The price of rope is $5.25 per foot. What is the price for 2.4 feet of rope?

Level 3: 15 feet of rope sells for $20.25. How much would 21 feet of rope cost?

Genius Level: If the price of 6.875 feet of golden rope is $990, what does 7 inches of golden rope cost?

Problem Set 2

Warmup: What is a better deal, 6 apples for $2 or buying 6 apples for $.35 each?

Level 1: What is the cost of 12 pencils if the price of each pencil is 24 cents?

Level 2: If 12 pounds of chicken cost $40.20, what would be the cost of 10 pounds of chicken?

Level 3: If the cost of electricity is 8.7452 cents per kilowatt, the 8 stands for 8 cents and the 7 stands for $7/10$ of a cent. What does the 2 stand for?

Genius Level: A 50 gallon container of milk leaks .2 gallons every 7.5 minutes. If the value of the milk is $4.25 per gallon, what is the value of the amount of milk that leaks in 2 hours and 15 minutes?

Problem Set 3

Warmup: If a bug walks .5 miles each hour, how long will it take the bug to go 3 miles?

Level 1: If it takes a dog .25 hours to walk one mile, how far will it walk in 3.25 hours?

Level 2: If a coin is .125 inches thick, how many will fit in a container that is 2.125 inches tall?

Level 3: Janet runs 4.25 miles every hour. How far will she run in 4 hours and 12 minutes?

Genius Level: David is planning a 6-day hike of 131.25 miles. He is planning to walk 6.25 hours each day of the trip. How many miles will he travel each hour of his trip?

Problem Set 4

Warmup: Write 99.9 in expanded form.

Level 1: In the number 83.125 the 1 stands for $\frac{1}{10}$ and the 2 stands for $\frac{2}{100}$. What does the 5 stand for?

Level 2: Write the base 8 number 64.25 in expanded form.

Level 3: Translate the base 2 number 1,000,000.01 into base 10.

Genius Level: Bill has $21.40 in base 10. Stanley has $21.40 in base 5. Who has more money? How much more?

Problem Set 5

Warmup: If a car can go 22.25 miles on one gallon of gas, how far will it go with 3 gallons of gas?

Level 1: A car can travel 28.5 miles on one gallon of gas. If the tank holds 19.5 gallons, how far can the car go with a full tank of gas?

Level 2: If the price of gas is $3.75 per gallon, how many gallons can you buy for $57.75?

Level 3: What is the cost of gas to drive 1530 miles if the car gets 25.5 miles per gallon of gas and gas cost $3.99 per gallon?

Genius Level: Car A is a hybrid car that gets 50 miles per gallon. Car B gets 40 miles per gallon. If gas cost $4.25 per gallon, how much more expensive is the gas to drive Car B compared to Car A if the distance traveled is 3000 miles?

Problem Set 6

Warmup: Jillian is boiling water to cook rice. The temperature of the water is 210.5°F. How many degrees must the temperature increase to reach the boiling temperature of 212°F?

Level 1: If the air temperature is 48.6°F, how many degrees must it drop to reach freezing? (32°F)

Level 2: If you have a fever and your temperature is 102.8°F, how far will your temperature have to drop to reach a normal body temperature? *(research needed)*

Level 3: Liquid oxygen will freeze at -368.77°F and will boil at -297.33°F. If you have frozen oxygen at -368.77°F, how many degrees must the temperature rise for it to boil?

Genius Level: Anastasia's refrigerator's temperature is 40.7°F. She wants to try and get the temperature to drop to absolute zero. How many degrees must the refrigerator's temperature drop until it reaches absolute zero? (Round to the nearest whole number.) *(research needed)*

Decimals
Level 1

1) A square that has a length of 2.5 inches on each side is folded in half to make a rectangle. What is the perimeter of the rectangle?

2) If the price of gas is $3.75 per gallon, what is the cost of 10 gallons?

3) A turtle is riding a bike at a speed of 12 miles per hour. How far will it travel in 1.75 hours?

4) 346 means 300 + 40 + 6. What does 412.7 mean?

5) If gold chain sells for 37 cents per inch, how much does a foot of gold chain cost?

Decimals
Level 2

1) If the temperature dropped from freezing to -28.6°F, how many degrees did it drop?

2) If the price of gas is $3.75 per gallon, how much will 12.4 gallons of gas cost?

3) If the size of a gas tank is 16.5 gallons, what is the highest price gas can be and still keep the cost of filling the tank under $100?

4) 682 means 600 + 80 + 2   What does 439.89 mean?

5) The boiling point of water at sea level is 212°F, but on the top of Mt. Everest the high elevation causes water to boil at 159.8°F. How much lower is the boiling point of water on Mt. Everest compared to sea level?

Decimals
Level 3

1) If a fever of 103.2 drops 2.2 degrees and then rises 4.9 degrees, how many degrees must it drop to return to a normal body temperature?

2) Car A has an 18.4 gallon tank and Car B has a 16.6 gallon tank. How much more expensive is it to fill Car A's gas tank than Car B's gas tank if the price of gas is $3.85 per gallon?

3) 499 means 400 + 90 + 9   What does 6,307.038 mean?

4) If nutmeg is $9.60 per pound, what is the cost of .20 ounces?

5) The grocery store charges $1.89 for a quart of milk. Stanley paid $49.50 for 12.5 gallons of gas. How much more expensive per quart is milk when compared to gasoline?

Decimals
Genius Level

1) If a full 5 gallon container leaks .4 quarts of water every 5.5 minutes, how long until it is empty?

2) A one pound ball would weigh .16 pounds on the moon and .38 pounds on Mars. How much would a ball that weighs 9.92 pounds on the moon weigh on Mars?

3) 880 means 800 + 80   What does 333.333303 mean?

4) Tippy the turtle leaves home at 12:00 noon to travel to a friend's house that is 3.195 miles away. If he travels at a speed of .45 miles per hour, at what time will he arrive at his friend's house?

5) Mount Everest is considered the highest point on Earth. Its height is 29,035 feet above sea level. The lowest point is the Mariana Trench, which is 10.911 kilometers deep. What is the difference between the highest and lowest point on Earth? Express your answer in meters and round the answer to the nearest whole number. (One foot = .304 meters, 1000 meters = one kilometer)

Chapter Five
Money

I work at a store after school and sometimes customers do something that confuses me. If they buy something for $12.38, they will give me a 20-dollar bill and 13 cents. Why would they do that? Are they trying to be mean and confuse me?

I've had the same thing happen to me. If the customer didn't do that, the change would be 2 quarters, a dime, and 2 pennies. He probably didn't want that kind of change to carry around.

Here is a quick way to figure out how to give change when someone gives you extra money.

Subtract from the price

$12.38   Price
-   .13   **Extra change from customer**
$12.25   **New price**

Now the amount of change is easy to determine: $20 - $12.25 = $7.75

Do the following problems without pencil and paper or a calculator:

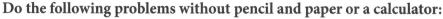

1) The cost of a goldfish is $7.87. The customer hands you a 10-dollar bill and 2 cents. What is the correct change?

2) Weightlifting ants cost $9.82 each. The customer hands you a 20-dollar bill and 7 cents. What is the amount of change?

3) 3 weightlifting ants cost $21.15. The customer hands you a 20-dollar bill, a 5-dollar bill and a quarter. What is the correct change?

I am going to Europe and I need to know how to change dollars into euros and euros into dollars.

One way to do this is to make two fractions. Let's say that one euro is equal to $1.50 and you want to know how many euros you will get for $30.

Fraction 1   Fraction 2

$$\frac{1 \text{ euro}}{\$1.50} \qquad \frac{? \text{ euros}}{\$30}$$

Now ask yourself what you multiplied by or divided by to get from $1.50 to $30. When you think a little, you can see that you multiplied by 20. The next step is to be fair to the top of the fraction and do the same thing — multiply by 20.

Fraction 1   Fraction 2

$$\frac{1 \text{ euro}}{\$1.50} \quad (\text{x } 20) \quad \frac{? \text{ euros}}{\$30}$$

Fraction 1   Fraction 2

$$\frac{1 \text{ euro}}{\$1.50} \quad \frac{(\text{x } 20)}{(\text{x } 20)} \quad \frac{20 \text{ euros}}{\$30}$$

1) If one euro is equal to $1.25, then how many euros would you receive for $100?

2) If 35 euros are equal to $50, how many euros are equal to $10?

3) If one euro is equal to $1.50, then what fraction of a euro is $1 equal to?

Problem Set 1

Warmup: If gold necklaces cost $90 per meter, what is the cost of 2½ meters of the necklace?

Level 1: If the price of a gold necklace is $90 per meter, what is the cost of 5 decimeters of necklace? *(Research needed)*

Level 2: If the price of a gold necklace is $90 per meter, what is the cost of 10 centimeters of necklace? *(Research needed)*

Level 3: If the price of a gold necklace is $90 per meter, what is the cost of 1 millimeter of necklace? *(Research needed)*

Genius Level: If the price of a gold necklace is $10,000 per meter, what is the cost of 1000 nanometers of gold necklace? *(Research needed)*

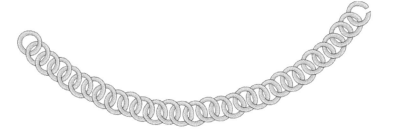

Problem Set 2

Warmup: If the price of three pounds of candy corn is $21, what is the cost for each pound?

Level 1: Ice cream is priced at 4 scoops for $5 or 3 scoops for $4.50. How much money do you save per scoop of ice cream when you buy 4 scoops?

Level 2: If a pound and a half of salmon cost $13.50, what does a pound of salmon cost?

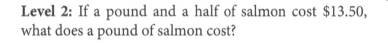

Level 3: A soccer ball and shin guards together cost $60. If the soccer ball is $10 more than the shin guards, what is the cost of the soccer ball?

Genius Level: The charge to rent a car is $80 per day plus 15 cents per mile driven. What is the cost to rent a car that is driven n days and 100 miles?

Problem Set 3

Warmup: If the price of gas is $4 per gallon, what does it cost to fill an empty 15 gallon gas tank?

Level 1: If the price of gas is $3.50 per gallon, what does it cost to fill an empty 20 gallon tank?

Level 2: If a car can go 25 miles on each gallon of gas, what will it cost to go on a 1000 mile trip if gas cost $4 per gallon?

Level 3: If the price of gas is $4.35 per gallon, how many miles would you be able to travel in a car that gets 23 miles per gallon if you had $87 to spend on gas?

Genius Level: Laura currently drives a pickup truck that gets 12 miles per gallon of gas. She is thinking of selling her truck and buying a Prius that gets 50 miles per gallon of gas. If the price of gas is $4 per gallon and Laura drives 15,000 miles in a year, how much money will she save on gas each year if she buys the Prius?

Problem Set 4

Warmup: If eggs cost $1.20 per dozen, what is the cost of one egg?

Level 1: If 3 dozen eggs cost $9.99, what is the cost of a dozen eggs?

Level 2: The cost of groceries was $18.37. If the clerk is given a 20-dollar bill and 2 cents, what is the change?

Level 3: What is the total cost of 99 shirts that cost $9.99 each?

Genius Level: A $100 phone is full price on December 1st. Each day the price will be cut in half. On which day in December will the cost of the $100 phone drop below $1?

Problem Set 5

Warmup: Sara spent $175 for 2 presents. If the cost of one present was $55, how much did the second present cost?

Level 1: The weight of the world's largest giant sequoia is 2,500,000 pounds. If you paid $2 per pound for the tree, what would it cost?

Level 2: If one euro is equal to 2 dollars, how many euros is one dollar equal to?

Level 3: A charge of .01 dollars per minute is how many times more expensive than a charge of .01 cents per minute?

Genius Level: If one euro is equal to $1.25, what fraction of a euro is one dollar equal to?

Money
Level 1

1) If the price of gas is $4 per gallon, what does it cost to fill a 22-gallon tank?

2) A 4000 pound van cost $24,000. What did the van cost per pound?

3) If 5 dozen eggs cost $15, what did each egg cost?

4) If a gallon of milk costs $4.80, what is the cost per quart?

5) It cost $63 to fill an empty 18 gallon gas tank. What was the cost of the gas per gallon?

Money
Level 2

1) The cost for a cup of coffee is $1.23. If the clerk is given $2 and 3 pennies, what is the change?

2) A type of hybrid car can travel 50 miles for each gallon of gas. If the price of gas is $4.50 per gallon, what would be the cost of gas for a 500-mile trip?

3) 1 ¼ pounds of bananas cost 75 cents. What did a pound of bananas cost?

4) If gold is worth $1850 per ounce, how much would a pound of gold be worth? *(Research needed)*

5) What is the value of a pile of money if it contains one of each bill from a $1 bill to a $100 bill? *(Research needed)*

Money
Level 3

1) If 3 ½ feet of rope cost 80 cents, how much did 14 feet of rope cost?

2) What is the total cost if Derrick bought a 99 cent pen, a book for $9.99, a bike for $99.99 and a computer for $999.99?

3) If a cell phone company charges .01 dollars per minute, how many minutes would have to be used before the charge was one dollar?

4) If a car can go 40 miles on a gallon of gas and the price of gas is $4 per gallon, what is the cost of gas per mile?

5) If gold costs $1850 per ounce, how much would a ton of gold cost? (*Research needed*)

Money
Genius Level

1) A Prius hybrid can travel 50 miles on a gallon of gas. A pickup truck can travel 15 miles on a gallon of gas. How much more will the cost of gas be for a 3000 mile trip if the pickup truck is driven instead of the Prius and the price of gas is $4 per gallon?

2) If one United States dollar is worth 80 cents in Canadian money, how many United States dollars is one dollar of Canadian money worth?

3) If a cell phone company charges .001 cents per minute, how many minutes would have to be used before the charge was one dollar?

4) Gold is selling for $1800 per ounce at Glen's Gold Shop. Keith's Gold By the Metric is selling gold for $99 a gram. Which store has the better price? Explain your answer. *(Research needed)*

5) A tank can travel .6 miles per gallon of gas. If gas is $3.75 per gallon, what would the cost of gas be to drive a tank from New York to Chicago? (The distance from New York to Chicago is 792 miles.)

Chapter Six
Fractions

The word fraction comes from a Latin word that means to break. We need fractions to measure "fractured" parts of a whole.

It would be great for math students if every measurement ended up being a whole number.

I had an uncle who didn't know how to use fractions and when he made cookies, he would round every number up to the nearest whole number. Look at what he did to my cookie recipe!

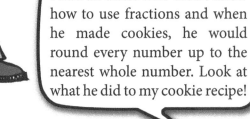

Best Cookie Recipe

Oats	2 cups
Baking powder	⅛ tablespoon
Banana	½ of whole
Sugar	⅜ cup
Applesauce	⅜ cup
Butter	⅝ stick
Raisins	1 cup

Best Cookie Recipe By My Uncle

Oats	2 cups
Baking powder	1 tablespoon
Banana	1 whole
Sugar	1 cup
Applesauce	1 cup
Butter	1 stick
Raisins	1 cup

Who wants cookies?

NO!!

I just ate!

Maybe next time if you learn to use fractions.

There are many different ways to work with fractions. The first way is called "Pizza and Pondering".

I get it! Pondering means thinking, so you are supposed to think about pizzas and fractions.

Problem 1: There are three pizzas and 9 people. How much pizza will each person receive? After drawing the pizzas, it is pretty easy to see that each person will get ⅓ pizza.

Problem 2: There are 4 pizzas and 7 people. How much pizza will each person receive? Because there are 7 people, divide each pizza into 7 pieces.

Each person receives ⅐ of each pizza for a total of 4/7 of a pizza. Fractions seem pretty easy!

Problem 3: What is ½ of ¼?

Another way of saying this is: What is ½ of ¼ of a pizza?

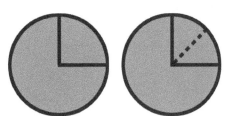

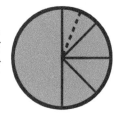

Now it is easy to see that ½ of ¼ = ⅛ of the pizza.

Problem 4: What is ½ of ¼ of ½?

Another way of saying this is: What is ½ of ¼ of ½ of a pizza?
First step: ¼ of ½.

When you divide the pizza in half, it is easy to see that ¼ of ½ of the pizza is equal to ⅛ of the pizza.

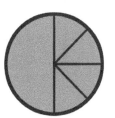

Now we need to find what is ½ of the ⅛ of the pizza. It is ¹⁄₁₆ of the pizza.

The next way to work with fractions is called *"Stupendous Shortcuts"*. This method still requires thinking, but it allows you to solve fraction problems much more quickly than when you draw pictures of pizzas. Let's do our 4 problems again using *"Stupendous Shortcuts"*.

Before we solve the problem, we need to look at the parts of a fraction.

Problem 1: There are three pizzas and 9 people. How much pizza will each person receive?

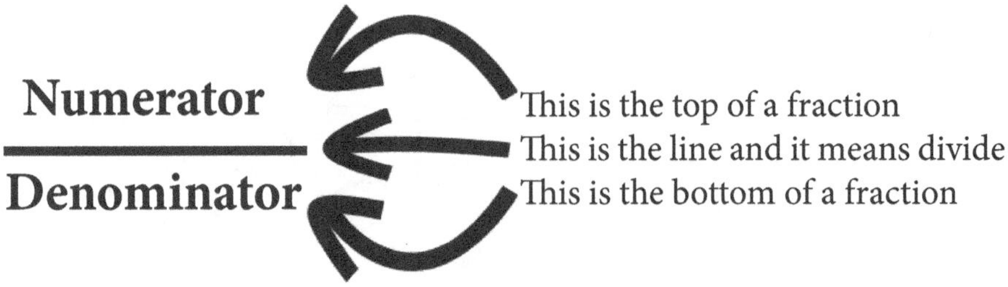

Numerator
———————
Denominator

This is the top of a fraction
This is the line and it means divide
This is the bottom of a fraction

This part is very, very important! A fraction means:

Numerator divided by denominator

$$\frac{\text{Numerator}}{\text{Denominator}} \quad \Longleftarrow \quad \text{Divided by}$$

$$\frac{\text{3 pizzas}}{\text{9 people}} \quad \Longleftarrow \quad \text{Divided by}$$

Let's turn the problem into a fraction. Because we are dividing the 3 pizzas, it goes on the top of the fraction.

That makes the problem very easy! Each person gets ³⁄₉ of a pizza. Because ³⁄₉ = ⅓, we can answer that each person will receive ⅓ of the pizza.

Problem 2: There are 4 pizzas and 7 people. How much pizza will each person receive?

$$\frac{\text{4 pizzas}}{\text{7 people}} \quad \Longleftarrow \quad \text{Divided by}$$

You're right -- It makes the problem very easy when you use the stupendous shortcut method! Each person gets ⁴⁄₇ of a pizza.

Problem 3 is a little different type of problem. Whenever you are taking a fraction of something, such as ½ of ¼, all you need to do is multiply the fractions. To do this, multiply the numerator by the numerator and the denominator by the denominator.

½ of ¼ means ½ x ¼ = ⅛

$$\frac{1}{2} \text{ x } \frac{1}{4} = \frac{1}{8}$$

Problem 4: What is ½ of ¼ of ½?

$$\frac{1}{2} \text{ x } \frac{1}{4} \text{ x } \frac{1}{2} = \frac{1}{16}$$

In the following problems, use both methods of fraction problem solving: *Pizza Pondering and Stupendous Shortcuts.*

1) Two pizzas will be shared by 9 people. How much pizza will each person receive?

2) What is ⅛ of ¼?

3) 4 people will be dividing ½ of a pizza. How much pizza will each receive?

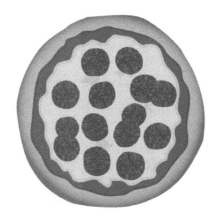

Problem Set 1

Warmup: There were 2 pizzas to feed 4 people. How much pizza would each person receive?

Level 1: There were 5 pizzas for 20 people. How much pizza would each person receive?

Level 2: There were 8 pizzas for 48 people. How much pizza does each person receive?

Level 3: There were 3 pizzas and 5 people. How much pizza would each person receive?

Genius Level: If there are 4½ pizzas for 36 people, how much pizza will each person receive?

Problem Set 2

Warmup: What is ½ of ½?

Level 1: What is ¼ of ½?

Level 2: What is ¼ of ¼?

Level 3: What is ½ of ½ of ½ of ½?

Genius Level: What is ½ of ¼ of ⅛?

Problem Set 3

Warmup: 2½ pies will be divided between 2 people. How much pie will each person receive?

Level 1: Two friends will be splitting ⅕ of a pie. How much of the pie will each friend receive?

Level 2: 3 friends will be splitting ½ of a pie. How much of the pie will each friend receive?

Level 3: Four mice decided to split ⅓ of a cheese wheel. What part of a whole cheese wheel will each mouse receive?

Genius Level: Five mice decided to split ⅔ of a cheese wheel. How much of a complete cheese wheel would each mouse receive?

Problem Set 4

Warmup: A mouse ate ⅞ of a piece of cheese. How much is left?

Level 1: Abe ate ¾ of a pie. His two friends split the rest. How much of the pie did each friend eat?

Level 2: Franklin's pet snail weighs 1 ⅜ ounces and his friend's snail weighs 1 ¹⁄₁₆ ounces. What is the weight of both snails?

Level 3: Dwight ate ½ of a pie while John ate ¼ of the pie. Lyndon ate ⅛ of the pie and Richard ate ¹⁄₁₆. How much of the pie is left?

Genius Level: Gerald ate ¹⁵⁄₁₆ of a candy bar and 3 friends split the rest. What fraction of the candy bar did each of the three eat?

Problem Set 5

Warmup: How much weight should be placed on the left side to balance the scale?

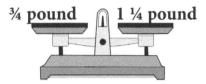

Level 1: How much weight should be placed on the left side to balance the scale?

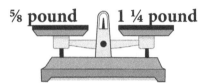

Level 2: How much weight should be placed on the left side to balance the scale?

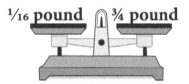

Level 3: How much weight should be placed on the left side to balance the scale?

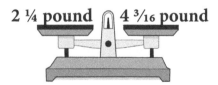

Genius Level: How much weight should be placed on the right side to balance the scale?

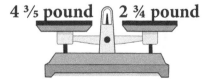

Problem Set 6

Warmup: A frog jumps ¼ foot each jump. How many jumps until it travels the 2 foot distance?

2 feet

Level 1: A frog jumps 1¼ feet each jump. How many jumps until it travels the entire 10 feet?

10 feet

Level 2: A frog travels 2½ feet each time it jumps. If it starts at the corner of a 5 foot by 15 foot rectangle and travels around the rectangle, how many jumps will it take before it travels around the entire rectangle?

15 feet

5 feet

Level 3: A frog travels 1⅛ feet each time it jumps on the short sides of the rectangle and 1¾ feet each jump on the long sides of the rectangle. How many jumps until it travels around the entire rectangle?

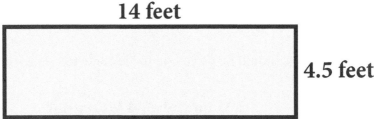

14 feet

4.5 feet

Genius Level: A frog must travel around a circle with a 21 foot circumference. Its first jump is 2¼ feet and the next jump is 2⅛. Because the frog gets tired each time it jumps, its jumps get shorter by ⅛ foot each jump. How many jumps until it travels completely around the circle?

Fractions

Level 1

1) There are 3 pies and 12 people. What fraction of a pie does each person receive?

2) What is ½ of ⅕?

3) Two friends are splitting ⅛ of a pie. How much of the pie do they each receive?

4) How much weight should be added to the right side to balance the scale?

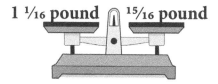

1 ¹⁄₁₆ pound   ¹⁵⁄₁₆ pound

5) A frog jumps ¾ of a foot each jump. How many jumps until it travels the entire 6 foot line?

6 feet

Fractions

Level 2

1) Laura found 32 plates that each had ¼ of a pie. If Laura put all the pieces of pie into whole pies, how many whole pies would she have?

2) ⅜ is how many times larger than ³⁄₁₆?

3) A meal with 50 calories has half the calories of a meal with 100 calories. A meal with 450 calories is what fraction of a meal with 3600 calories?

4) On a map, ¼ inch is equal to 12 miles. How many miles is ⅝ of an inch equal to?

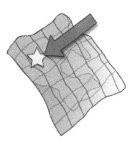

5) How many pounds should be added to the right side to balance the scale?

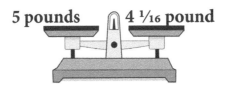

5 pounds   **4 ¹⁄₁₆ pound**

Fractions

Level 3

1) On Monday, Karen earned ½ the money she needed to buy a goat. On Tuesday, she earned ¼ the money she needs for the goat and on Wednesday she earned ⅛ what she needs. If the goat costs $96, how much more money does she need to earn to have enough money to buy the goat?

2) ⅓ of what fraction is equal to ¹⁄₁₅?

3) 2½ pizzas are going to be divided evenly among 20 people. How much pizza will each person receive?

4) How many **ounces** should be added to the left side of the scale to balance the scale?

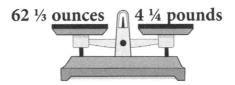

62 ⅓ ounces   4 ¼ pounds

5) Alicia found a bag of money. She gave ½ to charity and ⅓ to her brother. If she has 21 dollars left, how much money was in the bag at the start?

Fractions
Genius Level

1) If Lyn runs for 12 minutes each day, what fraction of the day is she running? (24 hours in one day)

2) How many pounds should be added to the left side to balance the scale?

3,000 kilograms   **4 ⅛ tons**

3) A person with $100 has ⅕ the money as a person with $500. A person with one penny has what fraction of the money of a person with $1,000,000?

4) What is ½ of ⅓ of ¼ of ½?

5) A super frog jumped half the length of a football field on his 1st jump and ⅓ the length on his second jump. Because he was tired, he jumped only ⅙ the length of the football field on his 3rd jump. How many more yards does the frog need to jump to get to the end of the football field?

Chapter Seven
Percents

Could you show me how to find 5% of $100?

To find percents, all you have to do is send things through the percent machine.

The percent machine cuts everything into 100 equal pieces. Let's send the $100 through the machine.

$100

Percent Machine

I see what happened. The percent machine split the $100 into 100 equal pieces. Each piece is $1, so 1% of $100 is $1.

$1 $1 $1 $1 $1 $1 $1 $1 $1 $1
$1 $1 $1 $1 $1 $1 $1 $1 $1 $1
$1 $1 $1 $1 $1 $1 $1 $1 $1 $1
$1 $1 $1 $1 $1 $1 $1 $1 $1 $1
$1 $1 $1 $1 $1 $1 $1 $1 $1 $1
$1 $1 $1 $1 $1 $1 $1 $1 $1 $1
$1 $1 $1 $1 $1 $1 $1 $1 $1 $1
$1 $1 $1 $1 $1 $1 $1 $1 $1 $1
$1 $1 $1 $1 $1 $1 $1 $1 $1 $1
$1 $1 $1 $1 $1 $1 $1 $1 $1 $1

If 1% is $1, then it is really easy to find 5%. 5% must be $5! Percent machines sure made that problem easy.

Now that you know how to use percent machines, tell me how to find 38% of this pie.

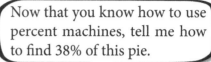

I am such a beautiful pie, don't put me through the percent machine!

Because the percent machine cuts things into 100 pieces, I think I should put the pie into the machine and then give you 38 of the pieces.

Now the pie is divided into 100 equal pieces. To find 38% of the pie, I simply take 38 of the pieces.

Percent Machine

If its okay with the worm, we can put it through the percent machine.

I'd be happy to go through the machine if it helps you understand percents a little better.

Percent Machine

I don't feel very good...

Now we have 100 equal size pieces of worm. If you want to give me 35½% of the worm, you need to give me 35 pieces plus half of another piece.

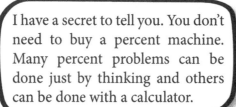

I have a secret to tell you. You don't need to buy a percent machine. Many percent problems can be done just by thinking and others can be done with a calculator.

What would I do if I wanted to find the sales tax for an $82 bike in a state with a 7% sales tax?

Just put the $82 through the percent machine or use a calculator and divide by 100. Each piece that comes out is 82 cents. There are 100 82¢ pieces. Now just take 7 of those pieces.

Percent Machine

After you've done a little work with percents, you'll soon find out that they are another way of saying a part of something. Look at the following parts of something.

25% of something is the same as saying ¼ of something.

50% of something is the same as saying ½ of something.

20% of something is the same as saying ⅕ of something.

100% of something is the same as saying all of it.

Here's one that is a little difficult to understand. If you say 200% of $50, you are talking about twice as much, or $100.

1) What is 1% of 200?

2) What is 25% of 1000?

3) What is 300% of 10?

4) What is 10% of 10?

5) What is 17% of 95?

6) What is 15% of 60?

7) What is 75% of 200?

8) What is 10½% of 80?

9) What is 20% of 20?

10) What is ½% of 500?

Changing a Percent to a Decimal

 The method we have been using to find a percent of a number requires thinking about percents. There is a shortcut that still requires a little thinking, but not as much. This method involves changing the percent to a decimal and then multiplying. Look at the percents below and the decimals I changed them to and see if you can figure out how I did it.

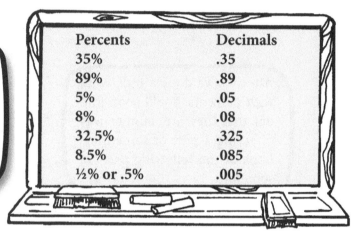

Percents	Decimals
35%	.35
89%	.89
5%	.05
8%	.08
32.5%	.325
8.5%	.085
½% or .5%	.005

I see what you did. You moved the decimal point two places to the left. Even though you don't see a decimal in 35% or 8%, there is one there. Just like with the numbers 35 and 8, there usually are not decimals showing, but they are there. 35. 8. 35.% 8.%

When 8% is written with a decimal, it is easy to move the decimal two places, but I have to put a zero in.

8.% — .08

To solve the problem 17% of 200, we can find 1% of 200. 200 ÷ 100 = 2. If 1% = 2, then 17% is 17 x 2 = 34.

We can also do the new shortcut and change 17% to a decimal and then multiply: .17 x 200 = 34.

1) A bike costs $175 plus sales tax of 6%. What is the total cost of the bike?

2) A $25 turkey was discounted 25% after Thanksgiving. What is the amount of the discount?

3) Ric had $480 in his bank account. If he spent 72.5% of his money, how much does he have left?

4) When people eat at restaurants, they usually leave a 15% tip. If the service was very good and you wanted to leave a 19% tip for a meal that cost $91, how much tip should you leave?

5) Brianna and David took Brianna's dog to the vet because it was sick. David offered to pay ¼% of the $100 vet bill. How much money is David offering to pay?

Comparing Numbers Using Percents

I just got my test back and I answered 64 questions correctly out of 80 questions. I wonder what my score is.

I know if there were 100 questions that your score would be 64%, but because there are only 80 questions, your score is different. I know you need to compare 64 to 80, but I am not sure how to do that.

It is very easy to compare numbers using percents. If you want to compare 64 to 80, the first thing you need to do is make a fraction. The line in a fraction means "compared to".

64 compared to 80 is written as $^{64}/_{80}$

This means 64 (compared to) 80

The line in a fraction also means divide so the answer is 64 ÷ 80 = .80. Now I need to change .80 to a percent. My guess is that you move the decimal two places to the right. I passed the test with a score of 80%.

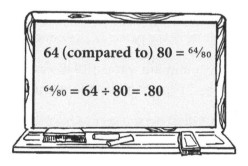

64 (compared to) 80 = $^{64}/_{80}$

$^{64}/_{80}$ **= 64 ÷ 80 = .80**

When we changed percents to decimals, we moved the decimal point two places to the left:

75.% = .75

If we want to change a decimal to a percent, we do the opposite, we move the decimal two places to the right:

.63 = 63%

Change the following decimals to percents:

1) .45   2) .08   3) .11   4) .005   5) 3.25

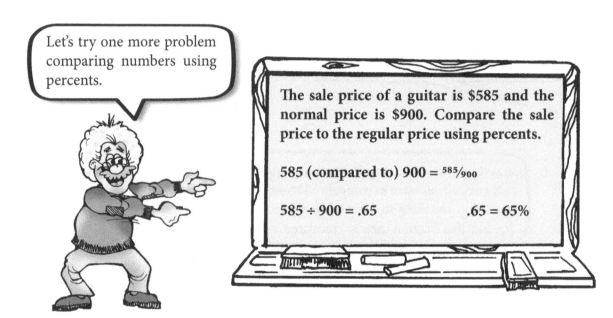

Let's try one more problem comparing numbers using percents.

The sale price of a guitar is $585 and the normal price is $900. Compare the sale price to the regular price using percents.

585 (compared to) 900 = $585/900$

585 ÷ 900 = .65   **.65 = 65%**

1) Ginny answered 45 questions correctly out of a total of 50 questions. What was her percentage score for the test?

2) Andrew missed 3 questions on a 30 question test. What was his percentage score for the test?

3) The Statue of Liberty is 305 feet tall. Compare a 4 foot tall child to the height of the Statue of Liberty using percents.   (Round to the nearest whole percent.)

4) The minimum wage in 1938 was 25 cents and in 2013 it was $7.25. The minimum wage in 1938 was __ percent of the minimum wage in 2013. (Round to the nearest whole percent.)

5) The tallest person ever to have lived was 8 feet 11 inches tall. The shortest person was 22 inches tall. The shortest person was __ percent of the tallest person.
(Round to the nearest whole percent.)

Percent of Increase or Decrease

There is another type of percent problem I don't understand. My mom said that she was getting $25 an hour and now receives $30 an hour. She said that was a 20% raise, but I am not sure how she came up with 20%.

I have trouble with those too! When gas went from $4 to $3, my mom said that the price decreased 25%. I am wondering why it is a 25% decrease.

To find the percent of increase or decrease, start by making a fraction. Put the amount something went up or down as the top of the fraction (numerator) and the original price or number as the bottom (denominator) of the fraction. I've solved the two problems on the board.

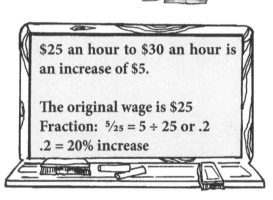

$25 an hour to $30 an hour is an increase of $5.

The original wage is $25
Fraction: ⁵⁄₂₅ = 5 ÷ 25 or .2
.2 = 20% increase

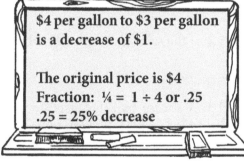

$4 per gallon to $3 per gallon is a decrease of $1.

The original price is $4
Fraction: ¼ = 1 ÷ 4 or .25
.25 = 25% decrease

1) What is the percent of increase when the price of an apple goes from 50 cents to 75 cents?

2) What is the percent of increase when a child grows from 4 feet to 5 feet?

3) What is the percent of decrease when an allowance drops from $4 per week to $1 per week?

4) What is the percent of decrease when a wage drops from $20 per hour to $18 per hour?

Problem Set 1

Warmup: If you wanted to leave a 15% tip for a $100 meal, how much would you leave for the waiter?

Level 1: Joseph left a 20% tip for a meal that cost $10. What is the amount of tip that Joseph left?

Level 2: What is a 15% tip for a $50 meal?

Level 3: Elizabeth left a $27 tip for a $90 meal. What percent tip did Elizabeth leave?

Genius Level: The cost of a meal plus a 15% tip was $92. How much of the $92 was the tip?

Problem Set 2

Warmup: How much sales tax would be paid for a bike that cost $100 if the sales tax rate is 8%?

Level 1: How much sales tax would be paid for a bike that cost $50 if the sales tax rate is 4%?

Level 2: If the sales tax rate is 7%, what is the amount of the tax for a $40 purchase?

Level 3: A soccer ball that used to cost $60 will be sold with a 20% discount on Saturday. If the sales tax rate is 5%, what will be the total cost of the soccer ball on Saturday? (Include sales tax)

Genius Level: Stan paid 7% sales tax for a book. If the sales tax Stan paid was $3.50, what was the price of the book?

Problem Set 3

Warmup: Abraham missed 6 questions on a 100 question test. What percent of the questions did he get correct?

Level 1: Jacob received a score of 80% on a math test that had 50 questions. How many questions did Jacob answer correctly?

Level 2: Jacob answered 48 questions correctly and 32 questions incorrectly. What percentage score did Jacob receive on his test?

Level 3: Jacob needs a score of at least 92% to receive an "A" in math class. If there are 64 questions on the test, how many does he need to get correct to score at least 92%?

Genius Level: Jacob needs an 87% score to get a B+ on a math test. If he gets a B+, he will be allowed to adopt a dog at the local animal shelter. Jacob answered 47 questions correctly and received a score of 87%. How many questions were on the test?

Problem Set 4

Warmup: A delivery company loses 1% of its packages each year. If it was supposed to deliver 100 packages, how many did it lose?

Level 1: A delivery company loses 3% of its packages each year. If it was supposed to deliver 1000 packages, how many did it lose?

Level 2: A delivery company loses 15% of its packages each year. If it was supposed to deliver 500 packages, how many did it lose?

Level 3: A delivery company lost 50 of 10,000 packages. What is the company's loss rate percentage?

Genius Level: If a company has a yearly loss rate of .0001%, how many packages are they expected to lose out of the 1,000,000 packages they are supposed to deliver each year?

Problem Set 5

Warmup: A coat that cost $300 is on sale for 50% off. What is the new price?

Level 1: An iPad is on sale for 10% off. If the regular price is $400, what is the new price?

Level 2: A soccer ball that normally cost $50 is on sale for 20% off. Lauren has a coupon that says to take an additional 25% off any sale item. What price did Lauren pay for the soccer ball?

Take an additional 25% off any sale item

Level 3: Mark saw the sign shown below and went to the store expecting to pay $50 for $100 running shoes that he wanted. When Mark brought the $100 shoes to the counter to pay, the clerk said that the discounted price was not $50. What was the new price?

All $100 running shoes are now 1/2% off!!

Genius Level: There is a weeklong sale on $800 laptop computers. Regular price is charged on Monday and then an additional 10% is taken off each day of the week. What is the price of the laptop computers on Friday?

Problem Set 6

Warmup: The price of a dog went from $100 to $101. What was the percent of increase?

Level 1:   Eric's allowance went from $10 per week to $15 per week. What was the percent of increase?

Level 2:   The value of Rachel's Roger Lemon baseball card went from $75 down to $15. What was the percent of decrease in the value of Rachel's baseball card?

Level 3: Ricardo's salary went from $15,000 per year in 2011 to $60,000 in 2012. What percent was his raise?

Genius Level: Tania's allowance was $40 per week when she was 7 years old. This is what happened to Tania's allowance over the next several years:

8 years old: Decreased 50%
9 years old: Increased 50%
10 years old: Decreased 50%
11 years old: Increased 50%
12 years old: Decreased 50%

What was Tania's allowance when she was 12 years old?

Problem Set 7

Warmup: Franklin put $100 in the bank and made $5 interest in a year. What percent was the interest rate at the bank where Franklin put his money?

Level 1: Sara put $400 in a saving account that paid 2% interest each year. How much interest will Sara earn in one year?

Level 2: Jamie put $800 in a savings account that pays 7% interest and Isaac put $800 in a savings account that pays 5% interest. How much more interest money will Jamie earn than Isaac in one year?

Level 3: Jordan put $1000 in a savings account that pays 3% interest. After one year the interest is added to his savings account. He now has $1000 + $30 = $1030 in his savings account. How much money will Jordan have in his saving account after one more year?
After 1 year: $1030
After 2 years: ?

Genius Level: An easy way to estimate how money in a savings account will grow is to find out how many years it will take for money to double. To do this, divide 72 by the interest rate.
For example: If you put $500 in a savings account that pays 6% interest, your money would double in 72 ÷ 6 = 12 years. In 12 years your $500 would grow to $1000 and in 12 more years the $1000 would grow to $2000.

If a baby was given $1000 when she was born and it was put in a savings account that paid 12% interest, how much money will the child have when she reaches a retirement age of 66?

Problem Set 8

Warmup: A worm is 5 inches long. If you want to use 50% of the worm for bait, how long of a piece will you use?

Level 1: Lindsay has $800 in her savings account. Lindsay spent 25% of her savings for an iPhone. What was the cost of the iPhone?

Level 2: The annual or yearly interest rate for Devon's $10,000 college loan is 5%. How much interest will Devon pay for this loan in one year?

Level 3: William was confused about how to find ½% of something. He thought that ½% of $1000 was equal to $500. What is ½% of $1000?

Genius Level: Matthew pays 22.9% annual interest on a $15,000 credit card bill. How much interest does Matthew pay each month for this credit card?

Problem Set 9

Warmup: Daniel decided to save 1% of his weekly allowance. If Daniel's allowance is one dollar per week, how much will he save each week?

Level 1: Amy's monthly allowance is 40% of Julie's allowance. If Julie's allowance is $80, what is Amy's allowance?

Level 2: Ripley's allowance is 150% of Stu's allowance. If Stu's allowance is $90, what is Ripley's allowance?

Level 3: Amber has 17½% of her allowance put into a college fund. If her allowance is $30 per month, how much money is put into her college fund each month?

Genius Level: Emily has Social Security tax and Medicare tax taken out of her paycheck each month. The Social Security tax is 6.2% and Medicare is 1.45%. If Emily's salary is $800 per month, how much is left after Social Security taxes and Medicare taxes are deducted?

Percents
Level 1

1) If sales tax is 5%, how much sales tax would be paid for the purchase of a $200 bike?

2) Jacob missed 12 questions on a test with 100 questions. What percent of the questions did Jacob get correct?

3) If a $50 iguana is discounted 20%, what is the sale price?

4) If Jon wants to leave a 15% tip for the waiter for a meal that cost $10, how much money should he leave for a tip?

5) Mike's allowance went from $8 per week to $12 per week. What percent of increase was Mike's new allowance?

Percents
Level 2

1) If a set of $50 earphones are discounted 5%, what is the new price?

2) If sales tax is 7%, how much sales tax would be paid for the purchase of a $250 dog?

3) Six people ate at a restaurant and their bill was $150. How much money should they leave for a tip if they want to tip 15%?

4) Matlin bought a $900 television and charged it to his credit card that has a yearly interest rate charge of 18%. How much interest will Matlin pay in a year if he leaves the $900 on his credit card?

5) Jasmine has $2000 in a savings account that earns 3% interest each year. How much interest will Jasmine earn in one year?

Percents

Level 3

1)   ½ of $100 is how much more than ½% of $100?

2)   A $200 coat now sells for $170. What percent is the coat discounted?

3)   Rachel started with $100. On Monday she spent 50% of the $100. On Tuesday she spent 50% of her remaining money and on Wednesday she also spent 50% of what was left. How much money does Rachel have now?

4)   Three people wanted to equally split a 15% tip for a meal that cost $125. How much money should each person leave for a tip?

5)   Eric has 15% of his weekly paycheck taken out for taxes. If the amount taken for taxes is $75, what is the total amount of Eric's paycheck before taxes are subtracted?

Percents
Genius Level

1) A jar has 530 blue marbles, 410 red marbles, 257 green marbles and 3 brown marbles. What percent of the marbles in the jar are brown marbles?

2) Nick needs to score 93% on a test with 87 questions. What is the highest number of questions Nick can get wrong and still score 93%?

3) If Dan charges $10,000 on a credit card at 20% annual interest, how much interest does he pay each day?

4) An $800 guitar is discounted 15%. What is the total cost of the guitar after a 7% sales tax is added?

5) A car that cost $2000 in 1976 cost $16,000 today. What is the percent of increase for the car?

Chapter Eight
Translating Decimals, Fractions, and Percents

When people talk about ½ of something, I hear them say 50%, .5 or ½. What is the correct way to say half of something?

I sometimes think that people are from different countries. Maybe there is a country called Percents and two others called Fraction and Decimal. People who are from the country of Percent use percents to talk about parts of things and people from the country of Fraction use fractions to talk about parts of things.

No, there are not different countries. Percents, decimals and fractions are all used to talk about parts of things. For example, you can say ¼, .25 or 25% when you are talking about a quarter of something. Each one is correct. Look at this chart to see how to translate percents, decimals and fractions.

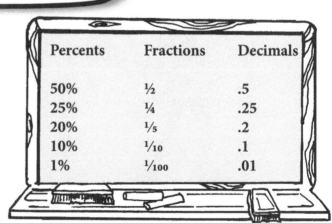

Percents	Fractions	Decimals
50%	½	.5
25%	¼	.25
20%	⅕	.2
10%	¹/₁₀	.1
1%	¹/₁₀₀	.01

Those are all easy to translate just by thinking about them, but sometimes I will need to change uneven percents to decimals and uneven decimals to percents. Can you review how to do that.

I remember from the last chapter that you translate percents to decimals by moving the decimal point 2 places to the left. Look how I changed all these percents to decimals. Remember that even though a decimal isn't written in percents such as 45%, there really is a decimal there: 45.% *becomes* .45

I am going to write all those percents with the decimal showing so it is easier to see how you moved the decimal point two places to the left to change percents into decimals.

Percents	Decimals
71%	.71
125%	1.25
7%	.07
.5%	.005
1000%	10

Decimals showing

Percents	Decimals
71.%	.71
125.%	1.25
7.%	.07
.5%	.005
1000.%	10

Look how easy it is to change from decimals to percents. You just move the decimal two places to the right this time.

Decimals	Percents
.82	82%
6.25	625%
.03	3%
.001	.1%
5	500%

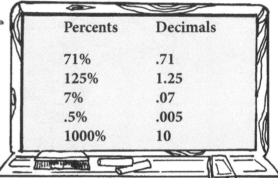

That was pretty easy, but what about changing a fraction such as ⅜ to a percent?

To change fractions to percents, you must first change the fraction into a decimal. Remember that the line in the fraction means divide so ⅜ is 3 ÷ 8.   3 ÷ 8 = .375

Now I know what to do. Move the decimal point 2 places to the right. The answer is 37.5%.

Look at how I change fractions to decimals and then to percents in the problems below. Some of the decimals are rounded to make it easier.

Fractions	Decimals	Percent
⅝	.625	62.5%
⅓	.333	33.3%
³/₇	.4286	42.86%
1 ⅞	1.875	187.5%
⁴/₉	.4444	44.44%

Translate the following into fractions, percents or decimals

	Percent	Fraction	Decimal
1)	35%	?	?
2)	?	¼	?
3)	?	¹/₂₀	?
4)	?	?	.01
5)	100%	?	?
6)	?	2 ½	?
7)	1000%	?	?
8)	?	⅛	?
9)	?	?	.09
10)	?	?	.085
11)	.25%	?	.0025
12)	?	⁷/₁₂	?
13)	2%	?	?
14)	?	?	.7
15)	26.4%	?	?

Problem Set 1

Warmup: 75% of the marbles in a bag are broken. What fraction of the marbles in the bag are not broken?

Level 1: A bag of marbles contains ½ green, ¼ red and the rest blue. What percent of the bag are blue marbles?

Level 2: 80% of a bag of marbles are red and the rest are green. What fraction of the bag are green marbles?

Level 3: ½ of a class received an A on a test, ⅕ scored a B , ¼ received a C and the rest failed the test. What percent of the class failed the test?

Genius Level: Boyd did ⅛ of his homework at school and 1/16 of his homework on the bus ride home. Boyd plans to do the rest at home. What percent of Boyd's homework does he plan to do at home?

Problem Set 2

Warmup: Frog A is ¼ the size of Frog B. Frog A is _____percent of Frog B.

Level 1: Building A is ⅕ the height of Building B. Building A's height is _____percent of Building B's height.

Level 2: A ruler's length is what fraction of a yardstick's length? A ruler's length is what percent of a yardstick's length?

Level 3: Building A is ⅕ the height of Building B. Building B is _____percent of Building A's height.

Genius Level: A millimeter is what fraction and percent of a meter?

Problem Set 3

Warmup: Mia's height is 50% of her mom's height. If Mia's mom's height is 6 feet 2 inches, what is Mia's height?

Level 1: Bill and Heath were comparing allowances. Bill found that his allowance was 100% of Heath's allowance. If Heath's allowance is $8 per week, what is Bill's allowance?

Level 2: Heath was comparing his allowance of $8 per week to his friend Jasmine's allowance. He found that Jasmine's allowance was 400% of his. What is Jasmine's allowance?

Level 3: Janet found that when she multiplied the amount of her allowance by 3¼ she could find her brother Thor's allowance. Thor's allowance is _____ percent of Janet's allowance.

Genius Level: Jasmine compared her $32 allowance to Isaac's allowance and found that it was 250% of Isaac's allowance. What is Isaac's allowance?

Problem Set 4

Warmup: If apples are 5 for $1, what is the cost of 3 apples?

Level 1: If 15% of the students at Einstein Elementary School are boys, what fraction of the students are boys?

Level 2: If Jamie has .67 dollars, how many cents does Jamie have?

Level 3: A store meant to sell 2 apples for 50 cents, but their sign read:

Based on this sign, what would be the cost of 4 apples?

Genius Level: In many states, a person driving with a blood alcohol level above .08% would be charged with drunk driving. What is .08% as a fraction?

Problem Set 5

Warmup: If a baseball player has 10 hits and 20 official appearances at the plate, what is his batting average?

Level 1: If a baseball player's batting average is .250, how many hits did he get if he had 100 official at bats?

Level 2: A baseball player's batting average is .200. How many hits did he get if he was at bat 225 times?

Level 3: If a baseball player has 205 hits and 505 official appearances at the plate, what is his batting average?

Genius Level: A National League pitcher has one hit in 165 at bats. What is his batting average?

Translating Decimals, Fractions and Percents
Level 1

Translate:

	Percents	Fractions	Decimals
1)	25%	?	?
2)	?	¾	?
3)	10%	?	?
4)	5%	?	?
5)	?	$\frac{1}{100}$	?

Level 2

Translate:

	Percents	Fractions	Decimals
1)	100%	?	?
2)	375%	?	?
3)	?	⅜	?
4)	1000%	?	?
5)	?	?	.875

Translating Decimals, Fractions and Percents
Level 3

Translate:

	Percents	Fractions	Decimals
1)	?	?	.025
2)	107%	?	?
3)	?	?	1.3333
4)	6.25%	?	?
5)	?	$^{15}/_{16}$	?

Genius Level

Translate

	Percents	Fractions	Decimals
1)	¼%	?	?
2)	?	?	.007
3)	?	$^{1}/_{1000}$	?
4)	.55.5%	?	?
5)	What is the difference between 50% of $800 and ½% of $800?		
6)	.025%	?	?

Chapter Nine
Metric System
Temperature: Fahrenheit and Celsius

I just received a call from my brother who is camping in Banff and Jasper National Park. He said the temperature there is 25 degrees, but it is expressed in Celsius not Fahrenheit. He wanted to know if he should be hot or cold.

Because your brother is camping in Canada, the temperature is given in a different temperature language called Celsius. It is very different than Fahrenheit. For example, the boiling point of water is 100 degrees in the temperature language of Celsius and 212 degrees in the temperature language of Fahrenheit.

This is a great temperature for a bath.

100° Farenheit

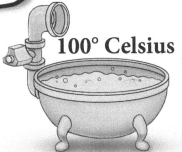

So if someone asked me if I was willing to take a bath in 100 degree water, I should make sure I ask if the 100 degrees is Celsius or Fahrenheit!

This bath is really, really hot. I feel like I am cooking myself. Are you sure it is only 100 degrees?

100° Celsius

 In the temperature language of Fahrenheit, the temperature of 100° Celsius is 212°. This means you would never want to swim in water that is 100° Celsius! There are machines on the next page that will change Celsius temperatures into Fahrenheit and Fahrenheit temperatures into Celsius.

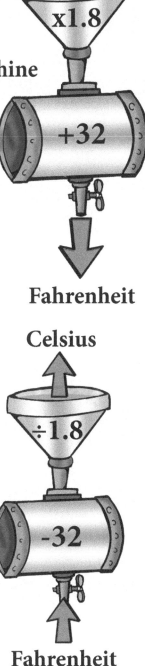

Celsius

x1.8

+32

Fahrenheit

I see now. Just like words are different in languages like Spanish, temperatures are different too. I know that the word for dog in Spanish is "perro". I am going to use the machines to change some temperatures from Fahrenheit into Celsius and from Celsius into Fahrenheit.

Celsius to Fahrenheit Machine

Those are really fancy machines. Let's try helping your brother by translating 25° in Celsius into Fahrenheit. I put the 25 into the Celsius to Fahrenheit machine:

25 x 1.8 = 45   45 + 32 = 77

Now we know that 25° Celsius is fairly warm because it is 77° F. Try the 5 problems shown below. Remember that when you are changing from Fahrenheit to Celsius, you go through the machine in the opposite direction and change addition and multiplication to the opposite operation.

Celsius

÷1.8

−32

Fahrenheit

Fahrenheit to Celsius machine. Go through machine in reverse and change to the opposite operation.

1) Change 5° Celsius to Fahrenheit.

2) Change 32° Fahrenheit to Celsius.

3) Change 72° Celsius to Fahrenheit.

4) Change 98.6° Fahrenheit to Celsius.

5) Change 0° Celsius into Fahrenheit.

Distance: Miles and Kilometers

When I drove to Banff and Jasper National Park, I crossed the border from Montana into Canada and saw speed limit signs that said I can drive 100 kilometers per hour. That seems a little fast!

We just learned about temperature languages. There are also different ways to say distance. The United States uses miles, but Canada and most other countries use kilometers.

My guess is that we are going to have to use a machine to change back and forth between kilometers and miles.

Here is the machine. Remember that if you need to change from kilometers to miles, just go through the machine in reverse and change to the opposite operation, which would be division.

Miles

x1.61

Kilometers

100 kilometers ÷ 1.61 = 62.1 miles. I guess 100 km/h is not that fast. I'm going to use the machine to try some of these problems.

1) The distance between New York and London, England is 5572 kilometers. How many miles is 5572 kilometers?

2) The distance from San Francisco to Honolulu is 2400 miles. How many kilometers is 2400 miles?

3) The distance to the sun is 93,000,000 miles. How far away is the sun expressed in kilometers? (Round to the nearest million.)

4) How many miles are in one kilometer?

5) The distance around the Earth is approximately 40,250 kilometers. What is the circumference of the Earth expressed in miles?

When you work with the metric system, you will also need to know about meters, decimeters, centimeters and millimeters. They are very simple to work with if you remember this information.

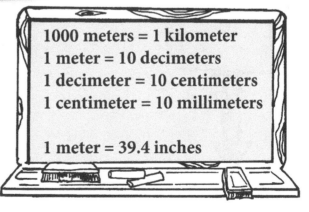

1000 meters = 1 kilometer
1 meter = 10 decimeters
1 decimeter = 10 centimeters
1 centimeter = 10 millimeters

1 meter = 39.4 inches

Try the following problems. I have also given you a meter to inches machine in case you need to change meters into inches, feet or yards.

1) How many decimeters are in 2.5 meters?

2) How many centimeters are in one meter?

3) How many millimeters are in one meter?

4) Approximately how many yards are in 2 meters?

5) How many inches are in 4 decimeters?

6) How many meters are in 197 inches?

7) How many millimeters are in 78.8 inches?

8) How many millimeters are in 3 decimeters?

9) 75.8 millimeters are equal to _____ meters.

10) How many millimeters are in 50 kilometers?

Meter

x39.4

Inches

Weights: Kilograms and Pounds

I also had another question about my trip to Canada. I weighed myself and the scale said I weighed only 25 pounds when I know that I weigh about 55 pounds. Why was that?

Earlier we talked about how distances were different because Canada uses the metric system. Weights also have their own "metric language". That scale was almost certainly telling you your weight in kilograms. The changing machine below shows you how to translate from kilograms to pounds.

So if I put the 25 through the machine, I find that my weight is 25 kilograms x 2.2 = 55 pounds. That is more like what I expected. I'm going to try a few different weights and put them through the changing machine.

Kilograms

x2.2

Pounds

1) 100 kilograms = _____ pounds.

2) 110 pounds = _____ kilograms.

3) Can a 20,000 kilogram truck cross a bridge with a weight limit of 40,000 pounds?

4) Mandy weighs 143 pounds. If she travels to Canada, what would she expect to weigh on a Canadian scale?

5) If gold is $24,000 per pound, what is its cost per kilogram?

Liters, Quarts and Gallons

When we need to measure volumes in the metric system we use liters and milliliters instead of gallons, quarts, pints and cups. This machine will help you translate from liters to quarts and also from quarts to liters. I've also put some information on the board that you will need to know.

When you try the problems below, remember that to convert from quarts to liters you need to go through the machine in reverse and change the operation from multiplication to division.

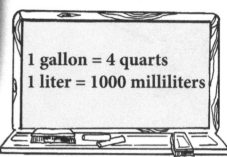

1 gallon = 4 quarts
1 liter = 1000 milliliters

Liters

x1.057

Quarts

1) 3.5 liters is equal to how many milliliters?

2) 8500 milliliters = _____ liters.

3) 4000 milliliters = _____ quarts.

4) 250 milliliters is equal to what fraction of a liter?

5) 1.057 quarts is equal to how many liters?

6) If you drank 4 liters of soda, you drank_____ quarts.

7) One gallon is equal to _____ liters.

8) What is more expensive, $3 per liter or $3 per quart?

9) If gas is $1.50 per liter, what is the cost for 4.228 quarts?

10) If gas cost $1 per liter, what is the cost of 50 quarts of gas?

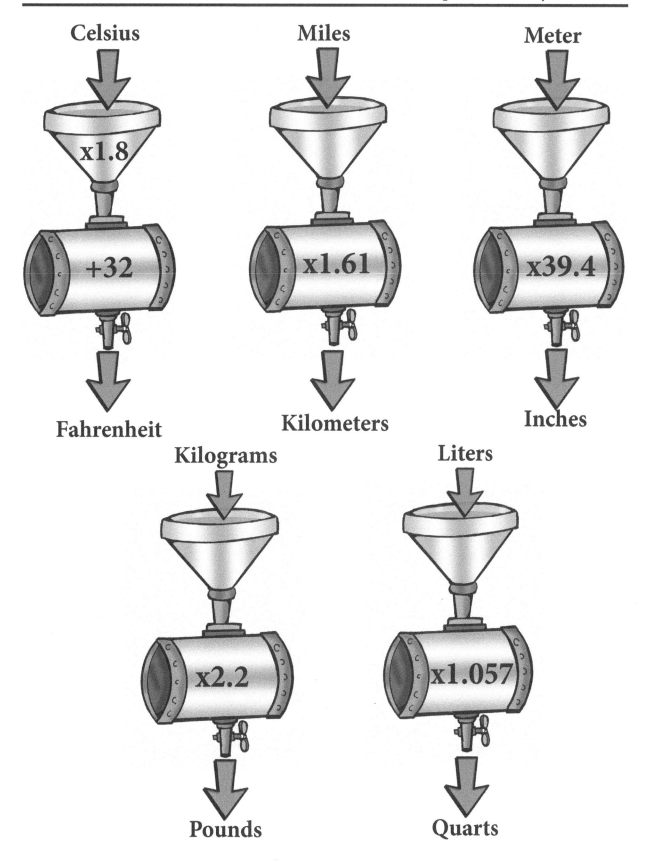

Problem Set 1

Warmup: Will water be frozen at 8° C?   Will water be boiling at 102° C? (at sea level)

Level 1: The normal body temperature is 98.6° F. What is this in Celsius?

Level 2: One day the temperature dropped 18 degrees from 68° F in the day to 50° F at night. How many Celsius degrees did the temperature drop?

Level 3: The coldest temperature possible is absolute zero because all molecular movement stops at that temperature. Absolute zero is -459.67° F. What is absolute zero in the Celsius scale?

Genius Level: 212° F is 100° in the Celsius scale. 32° F is 0° in the Celsius scale. There is one temperature where Fahrenheit and Celsius are exactly the same. What temperature are Fahrenheit and Celsius exactly the same? (Hint: it is very cold.)

Problem Set 2

Warmup: If you walk 10 miles, how many kilometers did you walk?

Level 1: A car was traveling at 41 mph. What was its speed in kilometers per hour? (Round to the nearest whole number.)

Level 2: The speed limit is 65 mph and a car is traveling at 100 kph. Is the car speeding?

Level 3: The distance from Chicago to New York City is 790 miles. If a car averages a speed of 82 kilometers per hour, how long will it take to travel from Chicago to New York City? (Round to the nearest half hour.)

Genius Level: The distance from New York City to Los Angeles is 2,790 miles. If you left New York City at 12.05 P.M. on September 7th and averaged a speed of 100 kilometers per hour, on what day and at what time would you arrive in Los Angeles?

Problem Set 3

Warmup: A child's swimming pool contains 1000 liters of water. How many quarts of water are in the pool?

Level 1: A quart is:

    a) Slightly larger than a liter
    b) Slightly smaller than a liter
    c) About 1.5 liters
    d) Equal to 4 liters

Level 2: Two gas stations are selling gas. Station A charges $1 per liter while Station B charges $4 per gallon. What station has the better price?

Level 3: If gas cost $4.20 per gallon, what does it cost to fill an 80 liter gas tank? (Round to the nearest dollar.)

Genius Level: An adult heart pumps approximately 1980 gallons of blood each day. How many liters of blood does an adult heart pump in a year? (Round to the nearest whole number.)

Problem Set 4

Warmup: A turkey weighs 3 kilograms. How many pounds is the turkey?

Level 1: Abe weighs 50 kilograms and Mary weighs 100 pounds. Who weighs more?

Level 2: A truck that weighs 22,000 pounds is about to cross a bridge with a weight limit written in kilograms. How many kilograms are in 22,000 pounds?

Level 3: There are 16 ounces in a pound. How many ounces are in a kilogram?

Genius Level: A store is selling an ounce of gold for $1500. A magazine is selling gold for $99 per gram. Is the magazine's gold a good deal? Explain your answer.

Problem Set 5

Warmup: The Statue of Liberty is 93 meters tall. How many feet tall is the Statue of Liberty? (Round to the nearest foot.)

Level 1: The Empire State Building is 1454 feet tall when the antenna spire is included. How many meters tall is the Empire State Building? (Round to the nearest whole meter.)

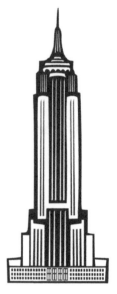

Level 2: The tallest building in the world is in Dubai and is 82,980 centimeters tall. How many meters tall is the building?

Level 3: The tallest building in the world's height (in the previous problem) is what fraction of a mile?

a) ¼ mile tall   b) ⅓ mile tall   c) ½ mile tall   d) ¾ of a mile tall

Genius Level: A picture of a large gold coin in a magazine has fine print that states that the actual size of the coin is 11.6 millimeters in diameter. The picture makes the size of the coin appear to be the size of a small plate. If you order the coin, when you receive it, the size will be very close to:

a) an aspirin   b) dime   c) quarter   d) small plate

Metric System
Level 1

1) A car was traveling at 55 miles per hour when the speed limit was 85 kilometers per hour. Was the car speeding?

2) One type of pterodactyl had a wingspan of 12 meters. How many feet is 12 meters? (Round to the nearest whole foot.)

3) The temperature of a hot tub is usually set at 104°F. What is this in Centigrade?

4) Who is running faster, runner A or runner B?

   Runner A: 5 miles per hour   Runner B: 8 kilometers per hour

   a) About the same speed   b) Runner B by a small amount
   c) Runner B is almost twice as fast as Runner A d) There is no way to tell

5) What are the boiling and freezing points of water using the Celsius scale? (at sea level)

Metric System
Level 2

1) A magazine ad was selling a gold coin. The picture of the coin took up half the page, but the small print mentioned that the diameter of the real coin was 24.3 millimeters. The actual size of the coin is closest in size to which item listed below?

     a) aspirin   b) dime   c) quarter   d) 4 inch wide pancake

2) How many square centimeters are in one square meter?

3) The speed of light is 186,000 miles per second. If you were writing a science book for Canadian students, the speed would need to be in kilometers per second. What is the speed of light in kilometers per second? (Round to the nearest 1000 kilometers per second.)

4) Approximately how many liters are in a 20 gallon gas tank?

     a) 20 liters   b) 45 liters   c) 75 liters   d) 100 liters

5) Which one of the following temperatures is closest to room temperature?

     a) 20˚C   b) 30˚C   c) 40˚C   d) 50˚C

Metric System
Level 3

1) The weight limit of a bridge is 40,000 pounds. A truck carrying cows that weigh 1500 pounds each weighs a total of 20,000 kilograms. How many cows must be taken off the truck before it can safely cross the bridge?

2) Mercury is a very unusual metal because it is liquid at room temperature. Mercury will freeze at -38.83°C. What is this temperature in Fahrenheit?

3) The heart pumps approximately 8640 liters of blood per day and pumps the entire amount of blood in the average body each minute. How many liters of blood does the average body contain?

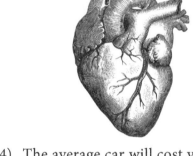

4) The average car will cost you approximately 55 cents per mile to drive. (This includes gas, repairs, insurance and the cost of the car) What is the cost per kilometer to drive a car? (Round to the nearest cent.)

5) How many pounds are in 200 grams?

   a) about 2 pounds   b) about 1.5 pounds   c) close to a pound   d) about ½ pound

Metric System
Genius Level

1) A truck that is hauling thousands of basketballs is crossing a bridge with a weight limit of 11 tons. If each basketball weighs 150 grams and the total weight of the truck and load is 10,150 kilograms, how many basketballs must be removed to bring the weight of the truck down to 11 tons?

2) Start at absolute zero expressed in Fahrenheit and round to the nearest whole number. Add the number of degrees Fahrenheit at which water boils. Add the number of degrees Fahrenheit that water freezes. Add the number of degrees Fahrenheit of a person's normal body temperature. Finally, add 17.4 to your answer. What number did you end up with?

3) At what speed in miles per hour, do you travel one kilometer per minute? (Round to the nearest whole number.)

4) The speed of the space shuttle was 27,870 kilometers per hour. How many miles per second is this speed? (Round to the nearest mile per second.)

5) A magazine ad for a gold coin said that the solid gold coin was a "full .98 grams". The cost of the gold coin was $195. If the price of gold is $1500 per ounce, is the gold coin a good buy? At $1500 per ounce, what should the gold coin cost? (grams x .035 = ounces)

Super Genius Level: A cubic meter is built entirely from cubic millimeter blocks. If all the cubic millimeter blocks are taken apart and put in a straight line, how long will the line be?

a) about 600 inches   b) about 600 feet   c) about 600 yards   d) about 600 miles

Chapter Ten
Language of Algebra

We just learned that the metric system is a different language for measuring things. Algebra is also a type of language -- It is a math language.

Some children think algebra is too hard for elementary age students because it is usually studied in high school. This is not true!! Algebra is actually very easy if you just remember that it is a math language. Watch how I translate the following math problems into the language of algebra.

Problem #1: Doug the dog is 3 years older than Celia the cat. If their ages total 17, how old is Celia the cat?

Language of algebra:
Celia the cat: n
Doug the dog: $n + 3$

When you are translating problems into the language of algebra, call the youngest animal n. Now we know that Doug the dog is 3 years older, so his age is $n + 3$.

Problem #2: Doug the dog is twice as heavy as Celia the cat. If their weights add up to 27 pounds, how much does Doug the dog weigh?

Language of algebra:
Celia the cat: n
Doug the dog: $2n$

We will call the animal that weighs the least n. Now we know that Doug the dog is twice as heavy, so his weight must be $2n$.

Try translating the following into the language of algebra. I have done the first two for you.

1) I ran a marathon in *n* hours. I ran the marathon in _____ minutes.
    The answer is 60n because there are 60 minutes for each hour.

2) The tree in my yard is *n* yards tall. The tree in my yard is _____ feet tall.
    The answer is 3n because for each yard there are 3 times as many feet.

3) It took me *n* minutes to wash the dishes. It took me _____ seconds to wash the dishes.

4) The hole in my back yard is *n* feet deep. The hole in my back yard is _____ inches deep.

5) Susan is 5 years older than Debra. If Debra is *n* years old, how old is Susan?

6) There are *n* gallons of water in my swimming pool. There are _____ quarts in my pool.

7) There are *n* dimes in a pile. The value of the pile expressed in cents is _____.

8) If there are *n* horses in a barn, how many horse legs are in the barn?

9) If there are *n* quarts of water in a lake, how many gallons are in the lake?

10) Bill is *n* years old. His sister is 3 times as old as Bill and his pet ferret is 5 years younger than Bill. What is Bill's sister's age and the age of his ferret in the language of algebra?

11) Avery bought *n* books and read 24 of them. How many books has Avery not read yet?

12) What is the total of 3 consecutive numbers if the smallest one is *n*?

13) Ian ran for *n* meters. How many centimeters did Ian run?

14) There are *n* pigs and twice as many ducks in a barn. How many legs are there in the barn?

15) A rectangle's length is twice its width. If the width is *n*, what is the perimeter of the rectangle?

There is a very easy way to translate things into the language of algebra if your brain seems to be confused.

I hope your suggestion will help because there are many problems that are hard to change into the language of algebra. My brain just starts to spin and can't figure out the way to do it! For example — How would I change *n* hours into seconds?

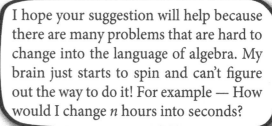

If you want to change *n* hours into seconds in the math language of algebra, simply ask yourself how to change 3 hours into seconds.

That would be easy. I would find out how many seconds there are in one hour. Let's see, there are 60 minutes in an hour and 60 seconds in each minute, so there would be 60 x 60 = 3600 seconds in each hour. The number of seconds in 3 hours would be 3 x 3600. That was easy!!

You just found that you can change any number of hours into seconds by multiplying by 3600. To change *n* hours into seconds, simply multiply by 3600!

n hours = $3600n$ seconds

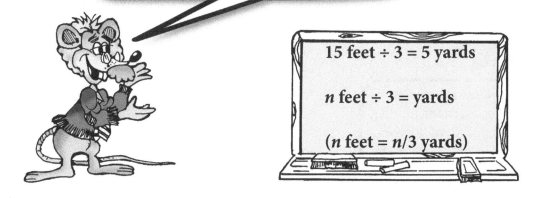

I am going to try this on another type of problem I have trouble with. If I have *n* feet, how many yards do I have? I will make up a number of feet, say 15 feet. Because there are 3 feet in each yard, it is easy to change 15 feet into yards and it is easy to change *n* feet into yards.

15 feet ÷ 3 = 5 yards

***n* feet ÷ 3 = yards**

(*n* feet = *n*/3 yards)

Change the following into the language of algebra. Put a number in for *n* to help your brain with the translation into the language of algebra.

1) How many yards are in *n* inches?

2) How many days are in *n* hours?

3) How many years are in *n* days?

4) How many gallons are in *n* pints?

5) How many hours are in *n* seconds?

Problems Set 1

Warmup: Ping has *n* legs. If Spot has 2 more legs than Ping, how many legs does Spot have?

Level 1: There are *n* cows in a field. How many legs are there?

Level 2: There are a total of 80 cows and ducks in a barn. If there are *n* cows, how many ducks are there?

Level 3: There are twice as many pigs as horses on a farm and a 5-legged dog named Extra. If the number of horses is *n*, how many legs are on the farm?

Genius Level: There are 100 total pigs and ducks on a farm. If there are *n* pigs, how many duck legs are there?

Problems Set 2

Warmup: Jacob has three more quarters than Blake. If Blake has *n* quarters, how many quarters does Jacob have?

Level 1: A jar of coins has 5 times as many dimes as quarters. If there are *n* quarters, how many dimes are there?

Level 2: A jar of coins has 5 times as many quarters as dimes. If there are *n* quarters, how many dimes are there?

Level 3: A jar of coins has *n* nickels. There are 7 times as many quarters as nickels in the jar. What is the value of the quarters? (In cents)

Genius Level: A jar of coins has 13 pennies and *n* nickels. There are twice as many dimes as nickels and twice as many quarters as dimes in the jar. What is the value of the coins in the jar?

Problems Set 3

Warmup: There are 2 consecutive numbers. If the smallest is called *n*, what is the other number called?

Level 1: There are 5 consecutive numbers and the smallest one is called *n*. What is the largest number called?

Level 2: There are 5 consecutive even numbers and the smallest is called *n*. What is the largest number called?

Level 3: What is the sum of 6 consecutive odd numbers if the smallest one is called *n*?

Genius Level: What is the average (mean) of 7 consecutive numbers if the smallest one is called *n*?

Problems Set 4

Warmup: Five yards are equal to 5 x 3 = 15 feet. Six yards are equal to 6 x 3 = 18 feet. *n* yards are equal to how many feet?

Level 1: If you walked *n* feet, how many inches did you walk?

Level 2: If you walk at a speed of 5 miles per hour, how many miles will you walk in *n* hours?

Level 3: How many seconds are in *n* hours?

Genius Level: If a slug crawls *n* feet, how many miles did it crawl?

Problems Set 5

Warmup: The length of a rectangle is 2 inches longer than the width. If the width is *n*, what is the length?

Level 1: The length of a rectangle is 4 times its width. If the width is *n*, what is the length?

Level 2: The length of a rectangle is 5 inches longer than 3 times the width of the rectangle. If the width is *n*, what is the length of the rectangle?

Level 3: The length of a rectangle is 8 times the width. If the length is called *n*, what is the width?

Genius Level: The length of a rectangle is 4 times its width. If the width is *n*, what is the perimeter of the rectangle? What is the area of the rectangle?

Problems Set 6

Warmup: Sandra is 3 years older than Danny. If Danny is *n* years old, how old is Sandra?

Level 1: Sandra is 3 years older than Danny. If Sandra is *n* years old, how old is Danny?

Level 2: Jon is 3 years younger than 6 times Gizmo's age. If Gizmo's age is *n*, how old is Jon?

Level 3: Alicia weighs 20 pounds more than Dan, who weighs 3 times as much as Bill. If Bill's weight is *n*, how much does Alicia weigh?

Genius Level: Ed is 4 times his grandson's age and 28 years older than his son. Ed's son is 6 years older than Ed's daughter. If their ages add up to 133, how old is Ed?

Language of Algebra
Level 1

1) If there are *n* chickens on a farm, how many chicken legs are there?

2) Sara is 32 years younger than her mother. If the age of Sara's mother is *n*, how old is Sara?

3) Trent had 3 quarters in his hand which are worth 75 cents. If Jillian has *n* quarters in her hand, how much are her quarters worth?

4) The length of a rectangle is 5 inches longer than its width. If the width is *n*, what is the length?

5) James has *n* spiders and 2 grasshoppers. How many legs does his insect collection have?

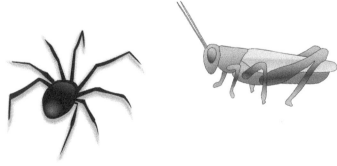

Language of Algebra
Level 2

1) There are 3 consecutive even numbers. If the smallest number is n, what is the largest number?

2) If a car travels at a speed of 60 miles per hour, how many miles will it travel in n hours?

3) If there are n gallons of water in a bucket, how many quarts of water are in the bucket?

4) A farm has n pigs, 20 cats and a 3-legged dog named Triangle. How many legs are there on the farm?

5) A snail crawled n meters. How many centimeters did it crawl?

Language of Algebra
Level 3

1) What is the sum of 5 consecutive numbers if the smallest one is called n?

2) There are a total of 1800 cows and ostriches on a farm. If there are n cows, how many ostriches are there?

3) The length of a rectangle is 9 times its width. If the width is n, what is the perimeter of the rectangle?

4) If there are n shoes in a closet, how many pairs of shoes are there?

5) Luke is half his father's age and twice his dog Shadow's age. Luke's grandfather is 48 years older than Luke. If Luke's age is n, write the age of Shadow, Luke's father, and Luke's grandfather in the language of algebra.

   Shadow:
   Luke: n
   Father:
   Grandfather:

Language of Algebra
Genius Level

1) There are *n* spiders and *n* chickens in a room. In the room, there are twice as many grasshoppers as spiders and there are 3 cows. How many legs are in the room?

2) It took Edwardo *n* hours to walk 10 miles. How many seconds did it take Edwardo to walk 10 miles?

3) There are a total of 200 buffalo and kangaroos at a ranch. If there are *n* kangaroos, how many buffalo legs are there?

4) The length of a rectangle is 2 inches shorter than 5 times its width. If the width is *n*, what is the perimeter of the rectangle?

5) Raymond is 10 times his grandson's age. Raymond's son is half his age and his daughter is 5 years younger than his son. If you call the grandson *n*, what is the sum of their ages?

Chapter Eleven
Algebra Problems

Now that you know how to change problems into the language of algebra, I am going to show you how to solve algebra problems.

Look at this very confusing problem that most people use guess and check to solve. It is a very easy problem if you change the problem into the language of algebra and then use algebra to solve it.

Five consecutive numbers add up to 185. What is the smallest number?

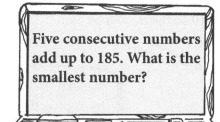

I remember how to change this into the language of algebra. I will call the smallest number *n*. The other 4 numbers are easy because you add one each time.

Language of Algebra

Smallest: n
Next: $n + 1$
Next: $n + 2$
Next: $n + 3$
Next: $n + 4$

Now you simply add up all 5 numbers and set them equal to 185. This is called making an equation. Using the word equation sounds like really advanced math, but it is really not too difficult.

Equation

$$n + n + 1 + n + 2 + n + 3 + n + 4 = 185$$

Now I have to add up the *n*'s and the numbers. I have $5n + 10 = 185$. What should I do next?

Equation

$5n + 10 = 185$

Equation

$5n + 10 = 185$
$ -10 \quad -10$
$5n = 175$

When you are solving equations, you always want to get the *n*'s alone on one side of the equation. We need to get rid of the 10 so the *n*'s will be alone.

We can easily get rid of the 10 by subtracting 10 from the left side of the equation. Algebra rules say we are allowed to do this as long as we are fair to the right side of the equation and do the exact same thing.

This is a very important rule to remember about algebra. You can do whatever you want to one side of the equation, as long as you are fair and do the EXACT same thing to the other side.

Now we have $5n = 175$. This means that 5 times some number = 175. I can find that number by dividing 175 by 5.
$175 \div 5 = 35$
The smallest number is 35! That was a lot easier than guess and check.

Let's try another problem. This one is pretty tricky, but I bet algebra will help to solve it. I know now that I first need to change the problem into the language of algebra and then into an equation which I need to solve.

A pile of money contains nickels and dimes. There are 5 times as many dimes as nickels and the value of the pile is $4.40. How many nickels are there?

Language of Algebra

Number of nickels: n
Number of dimes: $5n$

Value of the nickels: $5 \times n$ or $5n$
Value of the dimes: 10 times the number of dimes which is $5n$
        $10 \times 5n$ or $50n$

I see what you did. You needed to find the value of the nickels and the value of the dimes. You knew the number of nickels was n so you multiplied that by 5 to find its value. The number of dimes was $5n$ so you multiplied that by 10 to find its value. Now the equation is easy!

Equation

Value of nickels + value of dimes = 440 cents
$5n + 50n = 440$

So the first thing I need to do is add $5n + 50n$ which is $55n$. Now I know that 55 times some number = 440. The answer is $440 \div 55 = 8$. There are 8 nickels!

Equation

$5n + 50n = 440$
$55n = 440$

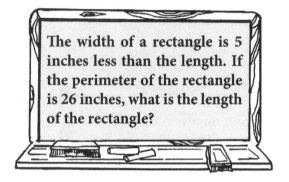

The width of a rectangle is 5 inches less than the length. If the perimeter of the rectangle is 26 inches, what is the length of the rectangle?

Here's another problem with a little twist. I want to make sure you know how to do it.

Language of Algebra

Length of rectangle: n
Width of rectangle: $n - 5$

Equation

$n + n + n - 5 + n - 5 = 26$
$4n - 10 = 26$

I see what you mean. I want the $4n$ to be alone, but if I subtract 10 from the -10, I will have -20. That doesn't help because I need to get the -10 to zero. I bet if I add 10 it will solve my problem!

Equation

$n + n + n - 5 + n - 5 = 26$
$4n - 10 = 26$
$+10 = +10$
$4n = 36$
$n = 9$

1) There are 5 more birds than snakes in a zoo. If the total number of snakes and birds is 25, how many snakes are there?

2) Brianna is twice as old as Brian. If the total of their ages is 21, how old is Brian?

3) There are 4 more nickels than dimes in a pocket. If there are only nickels and dimes and a total of 30 coins in the pocket, how many dimes are there?

4) Gabe is 19 years younger than Luke. If their ages add up to 23, how old is Gabe?

5) William is twice as old as William Junior and Grandpa is twice as old as William. If their ages add up to 140 years, how old is William?

The next 5 questions are somewhat difficult. Try them if you want a challenge:

6) There are twice as many ducks as cats on a farm and 8 cows. If cats, ducks and cows are the only animals on the farm and there are a total of 29 animals, how many cats are there?

7) There are only chickens, cows and a 6-legged dog named Hexagon on a farm. If there are three times as many chickens as cows and a total of 66 legs on the farm, how many cows are there?

8) Hiram is 4 times Dwight's age. If their ages add up to 105, how old is Dwight?

9) Dan is twice as old as Rachel and 3 years older than Bria. If their ages add up to 42, how old is Dan?

10) Five consecutive even numbers add up to 520. What is the smallest number?

Problem Set 1

Warmup: There are only pigs and a 3-legged dog named Tripod in a farm field. If there are 27 total legs, how many pigs are there? (Hint: call the number of pigs *n*)

Level 1: There are twice as many people as horses on a ship. If there are 135 total people and horses, how many horses are there?

Level 2: A petting zoo has only ducks and goats. There are 7 times as many ducks as goats and 54 total legs. How many goats are there?

Level 3: There are 126 heads in a barn of chickens, turkeys and pigs. If there are twice as many pigs as turkeys and twice as many turkeys as chickens, how many pigs are there?

Genius Level: There are only horses, ducks and a three legged dog named Triangle on a farm. There are 3 times as many horses as ducks and 213 total legs on the farm. How many ducks are there?

Problem Set 2

Warmup: There are 3 times as many quarters as nickels. If there are 16 total coins, how many nickels are there?

Level 1: There are twice as many nickels as pennies and twice as many dimes as nickels. If there are 77 total coins, how many pennies are there?

Level 2: There are two piles of dimes. The first pile has twice as many dimes as the second pile. If the value of both piles is 210 cents, how many dimes are in the smaller pile?

Level 3: A box contains only quarters and nickels. If there are 3 times as many quarters as nickels and the value of the coins is $4, how many nickels are there?

Genius Level: A box of coins contains nickels, dimes, quarters and half dollars. There are the same number of dimes as quarters. There are twice as many quarters as half dollars and there are the same number of nickels as half dollars. If the value of the box of coins is $18.75, how many quarters are there?

Problem Set 3

Warmup: Dave is a year older than Dan. If their ages total 21, how old is Dan?

Level 1: Mack is twice as old as Mitt. If their ages add up to 24, how old is Mitt?

Level 2: Stacy, Nicki and Lindsay are all the same age. If you add their ages and then add 25 to the total, you get the number 58. How old is Nicki?

Level 3: Ryan is 4 years older than Heather and Heather's age is 3 times Holly's age. If their ages add up to 60, how old is Ryan?

Genius Level: A tortoise in a zoo is 8 times older than the only bear in the zoo. The parrot is 17 years older than the bear and the rhino is 5 years younger than the parrot. If the ages of all four animals are combined, the total is 205 years. How old is the tortoise?

Problem Set 4

Warmup: Two consecutive numbers add up to 109. What is the smallest number?

Level 1: Three consecutive numbers add up to 153. What is the smallest number?

Level 2: There are five consecutive numbers that add up to 260. What are they?

Level 3: There are 4 consecutive odd numbers that add up to 176. What is the smallest number?

Genius Level: The average (mean) of 4 consecutive even numbers is 563. What is the smallest number?

Problem Set 5

Warmup: A square fence has a perimeter of 64 inches. What is the length of each side?

Level 1: A rectangle's length is twice its width. If the perimeter of the rectangle is 108 inches, what is its width?

Level 2: A rectangle's length is 3½ times its width. If the perimeter of the rectangle is 90 inches, what is the width of the rectangle?

Level 3: The length of a rectangle is 3 times its width. If the area of the rectangle is 432 square inches, what is the width of the rectangle?

Genius Level: There are two squares. The length of the side of the larger square is three times the length of a side of the smaller square. If the larger square's perimeter is 48 inches more than the smaller square's perimeter, what is the length of one side of the smaller square?

Algebra Problems
Level 1

1) Alicia's dog weighs 3 times what her pet ferret weighs. If their total weight is 10 pounds, what does the ferret weigh?

2) The life expectancy of a catfish is twice that of a bullfrog. If the life expectancy of a catfish and a bullfrog are added together, it would equal 90 years. What is the life expectancy of a catfish?

3) A pile of money contains quarters and nickels. If there are 4 times as many nickels as quarters, and there are 45 total coins, how many quarters are there?

4) Two consecutive numbers add up to 211. What are they?

5) A book and a bookmark together cost $5. If the book cost $4 more than the bookmark, what did the bookmark cost?

Algebra Problems
Level 2

1) The length of a rectangle is 6 ½ times its width. If the perimeter is 90 inches, what is the width?

2) Marissa is paid $12 per hour plus a $75 bonus each week. One week Marissa earned $495. How many hours did she work that week?

3) The weight of a killer whale is 8 times the weight of a polar bear, which is 11 times the weight of a chimp. If all three weights are added together, it would equal 10,000 pounds. What is the weight of the chimp?

4) Shadow is 5 years older than twice Ole's age. If their ages add up to 14, how old is Ole?

5) Mexico City's population is 5 million more than the population of Moscow, Russia. If the population of both cities adds up to 35 million, what is the population of Moscow?

Algebra Problems
Level 3

1) If 12 pounds are added to 3 bricks, the weight would be equal to the weight of 4 bricks with 12 pounds removed. What is the weight of each brick?

2) Bill weighs 40 pounds more than twice his dog's weight. If Bill and his dog weigh a total of 175 pounds, how much does Bill weigh?

3) Mike has $350 less than Kathleen. Jared has 4 times as much money as Mike. If all their money adds up to $5750, how much money does Mike have?

4) The life expectancy of a swan is twice that of a horse and 5 times that of a mountain lion. If the life expectancy of all three adds up to 170 years, what is the life expectancy of a mountain lion?

5) A newborn bison weighs 25 times more than a newborn polar bear. Newborn elephants weigh 5 times what newborn bison weigh. If the combined weight of all three newborns is 302 pounds, what does a newborn polar bear weigh?

Algebra Problems
Genius Level

1) Sara needs $90 to have enough money to buy a new iPad. Rachel needs $405 to have enough money to buy the same iPad. If they combine their money, they will have just enough money for the iPad. What does the iPad cost?

2) Luke's age is three times Daniel's age, which is half of Rachel's age. If their ages add up to 96, how old is Luke?

3) A box of money has 6 times as many dimes as nickels and 6 times as many dimes as pennies. The number of quarters is 3 times the number of nickels and the number of dollar bills is the same as the number of nickels. If the value of the money in the box is $12.05, how many quarters are there?

4) A farm contains horses, chickens and a 5-legged dog named Pentagon. If there are 4 times as many chickens as horses and there are a total of 185 legs, how many chickens are on the farm?

5) New York's population is twice the population of Paris.
Tokyo's population is twice the population of Los Angeles.
Tokyo's population is three times the population of Paris.

If the population of all 4 cities adds up to 75 million, what is the population of Paris?

Chapter Twelve
Probability

When we study probability, we are studying the likelihood of something happening. Here are some simple probability questions that are very easy to answer. We will start with these and then move on to some very interesting probability questions.

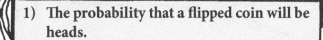

1) The probability that a flipped coin will be heads.

2) The probability that you will roll a "6" with one die.

3) The probability of picking a black marble when there are 10 black marbles in a jar of 50 marbles.

Those are so easy that they are boring.
½ or 1 in 2
⅙ or 1 in 6
¹⁰⁄₅₀ or 1 in 5

I'm ready for some challenging probability problems.

Okay, let's start with flipping coins. If you flip a coin twice, what is the probability that you will get 2 tails? How about flipping a coin 3 times and getting 3 heads? We can also use fancy math terms and instead of saying "what is the probability of getting 2 tails", we can write: P (tails, tails).

I see what you are doing with your fancy "math language". To solve the first problem, I need to find out what are all the possibilities if I flip a coin 2 times. I can get any of the following:

HH TT HT TH

Looking at all the possibilities, I can clearly see that of all 4, only one of the "outcomes" has both of the flips tails. The probability of flipping 2 tails therefore must be ¼!

The next problem is going to be a little more complicated. There are a lot of possibilities if a coin is flipped 3 times:

H,H,H	H,H,T
H,T,T	H,T,H
T,T,T	T,H,H
T,T,H	T,H,T

Now you have made it easy to find the answer. Of the 8 "outcomes", only one has all 3 heads. The answer must be ⅛. I can make another probability question from this information:

If you flip a coin 3 times, what is the probability that you will have at least 2 heads?

There are 4 of the "outcomes" that have at least 2 heads. The probability must be ⁴⁄₈ or ½.

H,H,H	**H,H,T**
H,T,T	**H,T,H**
T,T,T	**T,H,H**
T,T,H	T,H,T

As you flip the coin more and more times, it gets very time consuming to write out the number of outcomes. A shortcut helps a lot. Each flip of the coin doubles the number of outcomes:

2 flips: 4 events
3 flips: 8 events
4 flips: 16 events

If I flip a coin 30 times, I wonder if I will ever get 30 heads in a row. Every single flip must be heads so my probability is one out of however many outcomes there are. I wonder what the answer is if I multiply by 2 thirty times.

I looked on the internet to find out what the answer is if you multiply by 2 thirty times. It is a lot of outcomes!!

1,073,741,824

Your probability of getting 30 heads in a row is 1 in 1,073,741,824. Not very likely!

There is a box with 2 red marbles, one white marble, one green marble, one purple marble, and one black marble. If you close your eyes and pick 2 marbles from the box and put them in your pocket, what is the probability that both will be red?

In this type of problem, you will need to find the probability of the first marble being red and then the probability that the second pick is also red.

First pick red: ⅖ or ⅓

If the first is red, then there are 5 marbles left and only one is red.

Second pick red: ⅕

To find the probability of both events happening, simply multiply the probabilities. $\frac{1}{3} \times \frac{1}{5} = \frac{1}{15}$

So the probability of picking 2 marbles and both being red is $\frac{1}{15}$. Try the next problem. It is a little harder, but fairly easy if you find the probability of each thing happening.

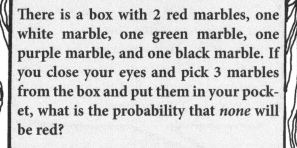

There is a box with 2 red marbles, one white marble, one green marble, one purple marble, and one black marble. If you close your eyes and pick 3 marbles from the box and put them in your pocket, what is the probability that *none* will be red?

First pick not red: 4 marbles are not red, so your probability of not picking a red marble is $\frac{4}{6}$ or $\frac{2}{3}$.

Second pick not red: If the first is not red, then there are 5 marbles left and 3 are not red. Probability of not red is $\frac{3}{5}$.

Third pick not red: If the first 2 picks are not red, then there are 4 marbles left and 2 are not red. Probability of not red is $\frac{2}{4}$ or $\frac{1}{2}$.

I'll use the same method as in the previous problem. To find the probability of all 3 picks not being red, I'll simply multiply the probabilities like before.
$\frac{2}{3} \times \frac{3}{5} \times \frac{1}{2} = \frac{6}{30}$ or $\frac{1}{5}$

The probability of picking 3 marbles and having them all not be red is only $\frac{1}{5}$.

I play the game of Risk and want to know what the probability is of rolling 3 sixes when I roll 3 dice. I think I'll use the same type of method you just used.

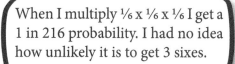

When I multiply ⅙ x ⅙ x ⅙ I get a 1 in 216 probability. I had no idea how unlikely it is to get 3 sixes.

> **First roll a 6: ⅙ or 1 in 6**
>
> **Second roll a 6: ⅙ or 1 in 6**
>
> **Third roll a 6: ⅙ or 1 in 6**
>
> **⅙ x ⅙ x ⅙ = ¹⁄₂₁₆**

1) What is the probability of flipping a coin five times and getting all heads?

2) Write problem one in "fancy math language".

3) There are 52 cards in a deck of cards. If you pick the ace of spades, you win. What is your probability of losing?

4) What is the probability of rolling 2 dice and getting 2 fives?

5) What is the probability of rolling 2 dice and getting the same number for each die?

6) Flipping a coin has two possible outcomes each time. When you roll a die, how many possible outcomes are there each time?

7) If you roll 2 dice, what is the probability that you will roll a seven?

8) If there are 100 families in a room and they all have 2 children, how many families would you predict would have both of their children boys?

9) There are 80 families in a room. They all have 3 children. How many of the 80 families would you predict would have all girls?

10) There are 100 marbles and only 2 are green. If you pick 2 marbles and put them in your pocket, what is the probability that they will both be green?

Problem Set 1

Warmup: If two coins are flipped, what is the probability that there will be 2 heads?

Level 1: If two coins are flipped, what is the probability that there will not be two tails?

Level 2: If three coins are flipped, how many possible outcomes are there?

Level 3: If a coin is flipped 7 times, how many possible outcomes are possible?

Genius Level: If a coin is flipped 10 times, what is the probability that there will be 10 heads?

Problem Set 2

Warmup: A box contains 3 red, 93 black and 4 white marbles. What is the probability of not picking a black marble?

Level 1: Julie has 10 quarters. She flips the first 9 and they all are heads. What is the probability that the 10th coin will be tails?

a) ½   b) Better than ½ probability because a tails flip is "due"   c) Not enough information

Level 2: There are two boxes that each have 10 black marbles and ten white marbles (a total of 20 marbles in each box). If Amy picks one marble from each box, there are four possible results she can have. The first possibility is that she will pick: (black marble, white marble) What are the other three possibilities that Amy might pick?

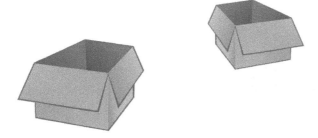

Level 3: There are 4 boxes that each have 50 white and 50 black marbles in them. If one marble is picked from each of the four boxes, what is the probability that all four would be white?

Genius Level: There is a box with 4 white and 6 black marbles in it. If Abe picks four marbles and puts them in his pocket, what is the probability that no black marbles would be picked?

Problem Set 3

Warmup: If you roll one die, what are you more likely to roll, a "1" or a "6"?

Level 1: If you roll one die, what is the probability of rolling an odd number?

Level 2: What is the probability of getting two of a kind if you roll two dice?

Level 3: When you roll two dice, there is only one way for the dice to add up to 2: 1 + 1
There are only two ways for the dice to add up to 3: 2 + 1 and 1 + 2. If you roll 2 dice, how many ways are there for the dice to add up to 7?

Genius Level: Stan and Jerry are each going to roll two dice. Stan needs to get two 5's to win a pet ferret. Jerry will win if he rolls any matching numbers. How much more likely is it that Jerry will win the ferret? (They can both win a ferret.)

    a) Each person has the same probability of winning
    b) Jerry is twice as likely to win
    c) Jerry is three times more likely to win
    d) Jerry is six times more likely to win
    e) Jerry is 10 times more likely to win

Problem Set 4

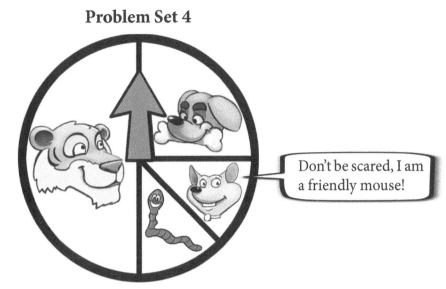

Don't be scared, I am a friendly mouse!

Warmup: If you spin the spinner, what is the probability that it will land on the dog section?

Level 1: If the spinner lands in the tiger area or the mouse area, you win a prize. What is your probability of winning the prize?

Level 2: Arial decided to play a game with her best friend where they would spin the spinner 10 times. The rules she made were that if the spinner lands in the tiger area, Ariel would collect $1. If the spinner lands in the mouse area, her friend would collect $4. Is this a fair game? Why?

Level 3: If the spinner is spun 8 times, predict how many times it will land in:
    Worm:
    Mouse:
    Tiger:
    Dog:

Genius Level: You have 8 tokens to place on the circle spinner. Each time the spinner lands in an area where you have tokens, you remove ONE and only one token. Whoever removes all their tokens first wins the game. For example, if you place all 8 tokens in the tiger area, then the spinner must land on tiger 8 times before you win the game.

You must place the tokens entirely inside one area and you can distribute the tokens however you want. You can place all 8 in one area or 2 in each area. What is the best way to distribute your tokens? Distribute the tokens however you want, but remember — If the pointer lands on an area where you have several tokens, you remove only one!

Problem Set 5

Warmup: There is a box with 2 black kittens, one white kitten, one spotted kitten, one gray kitten and one striped kitten. If you close your eyes and pick a kitten from the box, what is the probability of picking a black kitten?

Level 1: There is a box with 2 black kittens, one white kitten, one spotted kitten, one gray kitten and one striped kitten. If you close your eyes and pick a kitten from the box, what is the probability you will not pick a black kitten?

Level 2: There is a box with 2 black kittens, one white kitten, one spotted kitten, one gray kitten and one striped kitten. You close your eyes and pick a kitten from the box and it is not black. If you are given one more pick, what is the probability you will pick a black kitten? (After you pick, the kitten you pick is not put back in the box.)

Level 3: There is a box with 2 black kittens, one white kitten, one spotted kitten, one gray kitten and one striped kitten. If you close your eyes and pick two kittens from the box, what is the probability you will pick both black kittens? (After you pick, the kitten you pick is not put back in the box.)

Genius Level: There is a box with 2 black kittens, one white kitten, one spotted kitten, one gray kitten and one striped kitten. If you close your eyes and pick 3 kittens from the box, what is the probability you will not pick any black kittens? (After you pick, the kitten you pick is not put back in the box.)

Probability
Level 1

1) A class of 90 students are each told to pick a number from 1-10. A prize is won if the correct number is picked. Predict how many students picked the prize winning number.

2) If you need to roll one die and get a "5" to win, what is your probability of losing?

3) If the probability of getting struck by lightning is 1 in a million, what is the probability of not getting struck by lightning?

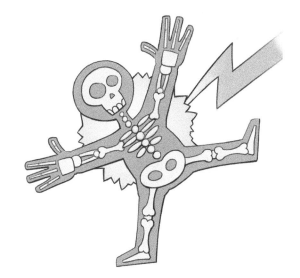

4) If the probability of getting the flu is 2 in 3, how many cases of flu would you predict for a class of 30 students?

5) There is a box with 5 green marbles, 20 red marbles, 25 black marbles and 50 yellow marbles. If you pick one marble, what is the probability of picking a green marble?

Probability
Level 2

1) If you pick one card out of a deck of 52 cards, what is the probability you will pick a spade?

2) If you pick one card out of a deck of 52 cards, what is the probability you will pick an ace?

3) If you pick one card out of a deck of 52 cards, what is the probability you will pick the ace of hearts?

4) You win a prize if you roll a "5" with one die or you can chose to flip two coins and get 2 heads to win the prize. Which method would you chose to try and win the prize, rolling the die or flipping the two coins? Why?

5) If you flip a coin 5 times, what is the probability you will get 5 heads in a row?

Probability

Level 3

1) If two dice are rolled, how many ways are there to get a roll of 10?

2) If you rolled one die two times, what is the probability that neither roll will be a "6"?

3) A jar of coins contains half dollars, quarters, dimes, nickels and pennies. The value of the half dollars is $5; the value of the quarters is $5; the value of the dimes, nickels and pennies are also each worth $5. If you picked one coin, what is the probability it would be a quarter?

4) What is the probability of rolling three "1"s when 3 dice are rolled?

5) Six ping pong balls are in a box. Each is labeled 1, 2, 3, 4, 5, or 6. Five balls are picked from the box and thrown away. What is the probability that the number 4 ball remains in the box?

Probability
Genius Level

1) There are 120 families in a room and each family has 2 children. How many of the 120 families would you predict have 2 boys?

2) There are 120 families in a room and each family has 2 children. How many of the 120 families would you predict have one boy and one girl?

3) Two dice are rolled once. If the total of the two dice add up to 6 or 8, a prize is awarded. What is the probability of winning a prize?

4) Four ping pong balls are in a box. They are each labeled 1, 2, 3, or 4. What is the probability that when they are taken out and thrown away that the order they are picked is 4 - 3 - 2 - 1?

5) If you picked 4 cards from a deck of cards, what is the probability that you will pick 4 aces?

Chapter Thirteen
Ratios

I know I am 4 feet tall and I just measured my shadow and found that it was 12 feet long. I wonder if that will help me find the height of this weird tree in my yard.

You can use ratios to solve many real life problems. Because you know that your shadow is 3 times your height, the tree's shadow must be 3 times its height.

I had no idea that it would be that easy to find the height of things. When I measure the tree's shadow, I found that it was 63 feet long. Now I know that the tree's height must be ⅓ of 63 or 21 feet.

That was an easy problem to figure out because it came out even. I want to show you a way to solve problems that do not come out even. It is called the cross-multiplying method.

Let's say that you are 3 feet tall and your shadow is 5 feet tall. You measure the tree's shadow and find that it is 60 feet. This makes things a little tricky because it is hard to do in your head. Watch how I set up this problem. (I put the tree's height as *n*, because that is what we are trying to find.)

$$\frac{3\text{ feet (Height)}}{5\text{ feet (Shadow)}} \;\; X \;\; \frac{n\text{ (Tree's height)}}{60\text{ feet (Tree's shadow)}}$$

The name of the problem solving technique is cross-multiplying so you simply multiply diagonally:

3 x 60 = 5 x *n*

I see what you are doing. 3 x 60 is 180, so 5 x *n* must equal 180. That's easy! 5 x 36 is equal to 180 so the tree must be 36 feet tall.

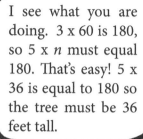

I think I can use this same method to work with maps and scales. If a scale says that one inch = 40 miles, it is very easy to see that 1.5 inches is equal to 60 miles, but when the distance between cities like Boston and New York is 5.2 inches, I have a hard time deciding what the distance is. I'll try cross-multiplying.

$$\frac{1 \text{ inch (Map)}}{40 \text{ miles (Real)}} \quad \mathbf{X} \quad \frac{5.2 \text{ inches (Map)}}{n \text{ miles (Real)}}$$

When I cross-multiply, I get 1*n* = 208. Now I know that the distance between Boston and New York is 208 miles. I can't believe how easy that was.

I am traveling from New York City to Seattle next week. On my large wall map with a scale of one inch = 40 miles, the distance is 73.75 inches. I'll use this fancy new method to find how far away Seattle is.

40 x 73.75 = 2950   so we know that:

***n* = 2950 miles.**

$$\frac{1 \text{ inch (Map)}}{40 \text{ miles (Real)}} \quad \mathbf{X} \quad \frac{73.75 \text{ inches (Map)}}{n \text{ miles (Real)}}$$

1) If a 5 foot post has a shadow of 30 feet, how tall is a person who has a shadow of 24 feet?

2) A 5 foot tall person has a shadow of 7.5 feet. How tall is a tree with a shadow of 90 feet?

3) A dog that is 3.25 feet tall has a shadow that is 4 feet long. A nearby building has a shadow of 176 feet. How tall is the building?

4) A map has a scale of 1.5 inches = 10 miles. How far apart are 2 cities that are 4.5 inches apart on the map?

5) A map has a scale of 1.5 inches = 10 miles. How far apart are 2 cities that are 12.75 inches apart on the map?

6) A globe has a scale of 1.75 inches = 100 miles. If two cities are 15.75 inches apart on the globe, how many miles apart are they?

There is a different type of ratio problem I want to show you. Say you know that a certain coin is made up of gold and silver in a weight ratio of 7 to 3. This means that for every 10 pounds of the coin, 7 of the pounds are gold and 3 are silver.

Look at this problem that appears to be difficult and watch how I solve it. Notice that the statue has 5 + 10 + 15 = 30 parts.

A statue is made up of copper, silver and gold in a weight ratio of 5:10:15. The total weight of the statue is 150 pounds. What is the weight of the gold in the statue?

There are 30 parts to the statue (5 + 10 + 15) and the parts weigh a total of 150 pounds, so each part weighs 150 ÷ 30 = 5 pounds. The gold has 15 parts so the weight of the gold in the statue is 15 parts x 5 pounds each = 75 pounds. See if you can solve this next problem.

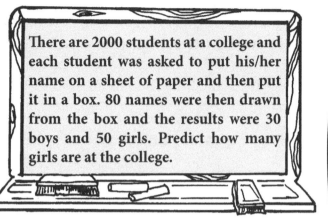

There are 2000 students at a college and each student was asked to put his/her name on a sheet of paper and then put it in a box. 80 names were then drawn from the box and the results were 30 boys and 50 girls. Predict how many girls are at the college.

I think that I can solve the problem just like the coin and statue problems. I will first find out how many groups of 80 students there are at the college of 2000 students. (I found the 80 in a group by adding 30 boys and 50 girls.)

2000 students ÷ 80 = 25 groups of 80 students

Each group has 50 girls:
25 groups x 50 girls per group = 1250 girls

Prediction: 1250 girls at the college

1) A coin is made up of nickel and copper in a weight ratio of 9:1. If the coin weighs 80 grams, what is the weight of the copper?

2) A statue is made up of silver, gold, and platinum in a weight ratio of 5:1:1. If the weight of the statue is 1050 pounds, what is the weight of the gold in the statue?

3) A school had 850 students. 50 of the students were chosen at random and there were 30 boys and 20 girls. Predict how many boys are in the school.

4) **(Genius Level)** A round ball is made up of gold and nickel in a weight ratio of 8:7. If the weight of the nickel in the ball is 56 pounds, what is the weight of the ball?

Problem Set 1

Warmup: Dell is 3 feet tall and has a shadow of 6 feet. A 4 foot tall person is standing next to Dell. How long is his shadow?

Level 1: At 3:00 in the afternoon, a 12 foot tall watchtower has a 3 foot shadow. A nearby tree has a 5 foot shadow. How tall is the tree?

Level 2: If a yardstick has a shadow of 6 inches, what is the shadow of a 12 inch ruler?

Level 3: A 10 foot tall tree has a shadow of one foot and 3 inches. A nearby post has a shadow of 6 inches. How tall is the post?

Genius Level: A yardstick has a shadow of 3.6 inches. How many centimeters long is the shadow of a meter stick?

Problem Set 2

Warmup: New York and Los Angeles are 3 inches apart on a map. If the scale of the map is one inch = 900 miles, how many miles is it from New York to Los Angeles?

Level 1: New York and Chicago are 4 inches apart on a map. If the scale of the map is ½ inch = 100 miles, how many miles is it from New York to Chicago?

Level 2: New York and Seattle are 11.25 inches apart on a large map. If the scale is 1 inch = 240 miles, how many miles is it from New York to Seattle?

Level 3: New York, New York and Anchorage, Alaska are 10 feet 5 inches apart on a large wall map. If the scale is 2½ inches = 90 miles, how many miles is it from New York to Anchorage?

Genius Level: The scale on a large globe is ¾ inch = 125 miles. If the circumference of the Earth is 25,000 miles, how many feet around is the circumference of the globe?

Problem Set 3

Warmup: A flea ½ inch tall can jump 2 inches high. If a child 5 feet tall had the ability to jump like a flea, how high could he jump?

Level 1: A flea that is ¼ inch tall can jump 5 inches high. If a child 5 feet tall had the ability to jump like a flea, how high could he jump?

Level 2: A flea that is ¼ inch tall can jump 8½ inches high. If a child 4 feet tall had the ability to jump like a flea, how high could he jump?

Level 3: A person 5 feet tall can jump 2½ feet high. If a flea that is ⅜ inches tall had the ability to jump like this person, how high could it jump?

Genius Level: A flea ¹⁄₁₆ of an inch tall can jump 5⅜ inches high. If a person 4 feet tall had the ability to jump like a flea, how high could he jump?

Problem Set 4

Warmup: A statue is made of gold and silver in a weight ratio of 5:3 (Gold to Silver). If the statue weighs 8 pounds, how many pounds of silver are there?

Level 1: A statue is made of gold and silver in a weight ratio of 5:3 (Gold to Silver). If there are 15 pounds of gold in the statue, what is the weight of the silver?

Level 2: A statue is made of gold and silver in a weight ratio of 5:3 (Gold to Silver).
If the total weight of the statue is 48 pounds, what is the weight of the gold in the statue?

Level 3: A statue is made of gold, silver and nickel in a weight ratio of 5:10:20. If the weight of the statue is 175 pounds, what is the weight of the nickel in the statue?

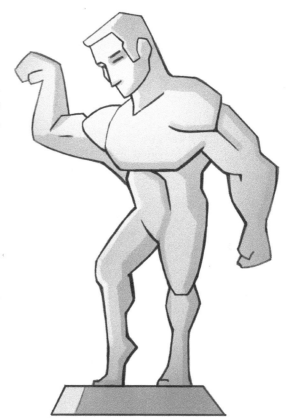

Genius Level: A statue is made of gold and silver in a weight ratio of 5:3 (Gold to Silver).
If the weight of the gold in the statue is 22.5 pounds, what is the weight of the statue?

Problem Set 5

Warmup: A class of 100 students all put their names on a piece of paper and put them in a hat. Laura picked 10 names and found that there were 5 boys and 5 girls. Predict how many boys are in the entire class of 100 students.

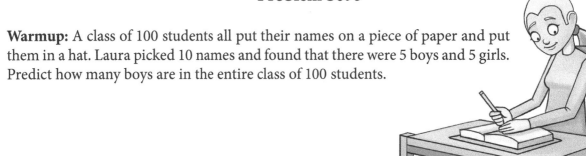

Level 1: A class of 30 students all put their names on a piece of paper and put them in a hat. Laura picked 10 names and found that there were 2 boys and 8 girls. Predict how many boys are in the entire class of 30 students.

Level 2: The third grade at Einstein Elementary School has 90 students. The first 15 third grade students to arrive at school were 10 boys and 5 girls. Predict how many girls there are in the third grade at Einstein Elementary School.

Level 3: There is a school with 1000 students. The name of each student was written on a piece of paper and all 1000 pieces were placed in a box. Stacey picked 100 names from the box and found that there were 36 boys and 64 girls. Using Stacey's information, predict how many boys are in the school.

Genius Level: There are 720 students at a school. 96 students' names were picked at random and the result was 40 boys and 56 girls. Predict how many girls are in the school.

Problem Set 6

Warmup: If a bag of dog food will feed 4 dogs for 2 weeks, how long will the bag feed 8 dogs?

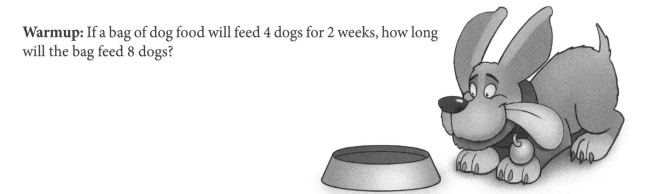

Level 1: If a box of cat food will feed 3 cats for 2 weeks, how long will it feed one cat?

Level 2: If 20 goldfish need 40 shakes of food in a week's time, how many shakes would 15 goldfish need for one week?

Level 3: If 12 pounds of food will feed a cat for 60 days, how many days will 12½ pounds of food last?

Genius Level: 8 people have enough food for a 6 day hike. If 4 more people join them, how many days will the food last?

Ratios
Level 1

1)   A 10-foot flagpole has a shadow of 2 feet. How tall is a tree with a shadow of 11 feet?

2)   A statue is made up of nickel and copper in a weight ratio of 8 : 5 (Nickel to copper). If there are 15 pounds of copper in the statue, how many pounds of nickel are there in the statue?

3)   A large globe has a scale of 2 inches = 1000 miles. If the distance on the globe from the North Pole to the South Pole is 25 inches, how many miles is it from the North Pole to the South Pole?

4)   A school asked all 100 students to write their name on a piece of paper and all names were put into a hat. 25 names were picked. Of the 25 names picked, there were 10 boys and 15 girls. Predict how many boys attend the school.

5)   A flea that is ½ inch tall can jump 12 inches high. If a child 5 feet tall had the ability to jump like a flea, how high could he jump?

Ratios
Level 2

1) A bug that weighs one ounce can carry 2 pounds. If a person who weighed 100 pounds had the ability to carry weight like the bug, how much weight could he carry?

2) An 80 pound statue is made up of tin and gold in a weight ratio of 7 : 3 (Tin : gold). How many pounds of gold are in the statue?

3) If a yardstick has a shadow of 6 inches, how tall is a tree with a shadow of 12 feet?

4) If 2 boxes of food will last 60 days for 2 cats, how long will 2 boxes of food last for 10 cats?

5) A flea that is ⅛ of an inch tall can jump 3½ inches high. If a 3 foot tall child had the ability to jump like a flea, how high could she jump?

Ratios
Level 3

1) A recipe for a cake calls for sugar, flour and oats in a ratio of 2:6:9. If there will be 68 total cups of sugar, flour and oats in the cake, how many cups of oats should there be?

2) The ratio of green to black marbles in a jar is 2 : 3. The green marbles were all counted and there were a total of 84 green marbles in the jar. How many black marbles are there in the jar?

3) A chainsaw requires a gas and oil mixture in a ratio of 32:1. If a gallon of gas is used, how many cups of oil should be added?

4) A flea that is ⅛ inch tall can jump 2½ inches high. If a 20 foot tall giraffe had the ability to jump like a flea, could it jump over the Statue of Liberty? (Including the foundation and pedestal) Defend your answer.

5) A school is made up of 250 boys and 350 girls. If you randomly picked 48 students' names from a list of students, predict how many boys would be picked.

Ratios

Genius Level

1) The scale on a map is ½ inch = 60 miles. If two cities are ¹⁵⁄₁₆ inches apart on the map, how many miles apart are they?

2) ¼ cup of ink is poured into a gallon of water and stirred for 5 minutes. A cup of the ink/water mixture is then taken out and put in a bowl. What is the ratio of ink to water in the bowl?

3) A school has 665 students. If the students are in a ratio of 4 girls to 3 boys, how many girls are in the school?

4) If a meter stick has a shadow of 60 centimeters, how tall is a tree that has a shadow of 75 feet?

5) A statue is made up of gold, silver and platinum in a weight ratio of 7:6:5. If the weight of the statue is 189 pounds, what is the weight of the platinum?

Chapter Fourteen
Measurement

When you work with measurements, it is very important that you know many facts about gallons, quarts, time, distances and weight. I've placed a lot of important information here to help you solve problems in this chapter.

3 teaspoons = 1 tablespoon
2 tablespoons = 1 fluid ounce
8 fluid ounces = 1 cup
2 cups = 1 pint
2 pints = 1 quart
4 quarts = 1 gallon

12 inches = 1 foot
3 feet = 1 yard
5280 feet = 1 mile
astronomical unit = 92,955,807 miles

light-year =
Approximately 6,000,000,000,000 miles

16 ounces = 1 pound
2000 pounds = 1 ton

Let's try a problem using volumes of water. Remember to be careful changing from gallons to quarts or between other sizes.

A sink that is badly in need of a plumber is draining at a rate of 15 gallons per hour. How many quarts are draining per minute?

I think the first thing I need to do is change the 15 gallons to quarts. 4 quarts per gallon x 15 gallons = 60 quarts. If 60 quarts are draining every hour, then one quart drains every minute because there are 60 minutes in one hour.

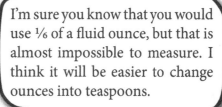

I always have trouble when I have to change a recipe. I have a recipe for 60 cookies that calls for one fluid ounce of vanilla. If I wanted to make 10 cookies, how much vanilla would I use?

I'm sure you know that you would use ⅙ of a fluid ounce, but that is almost impossible to measure. I think it will be easier to change ounces into teaspoons.

I know that one fluid ounce is equal to 6 teaspoons. Now it is pretty easy to find ⅙ of a fluid ounce. ⅙ of 6 teaspoons is one teaspoon.

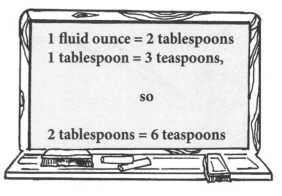

1 fluid ounce = 2 tablespoons
1 tablespoon = 3 teaspoons,

so

2 tablespoons = 6 teaspoons

1) If the price of milk is $5.20 per gallon, what does it cost for each pint?

2) Water comes out of a hose at a rate of 2 gallons per minute. How long does it take for a quart to come out of the hose?

3) If a snail crawls 16 feet per hour, how many days will it take for it to crawl a mile?

4) If sound travels 1100 feet per second, how many seconds does it take for sound to travel one mile? (Round to the nearest whole number.)

When we talk about very large distances, we cannot use feet or even miles because the distances are so large.

I know what you mean. When I talk about the former planet Pluto, I have to say that it is several billion miles away from the sun. That is hard to understand because the number is so large.

I was thinking about the star system Alpha Centauri the other day and wanted to tell a friend about it. He asked how far away it was and when I told him that it was about 25 trillion miles, he didn't really understand how far that is. He also didn't like to think about such a large number. There has to be a better way to express very long distances.

Fortunately there are a couple ways to express large distances. The first is called an astronomical unit. It is the distance from the Earth to the sun. It is about 93,000,000 miles. If you want the exact measure of an astronomical unit it is 92,955,807 miles.

Okay, let's see if I can use astronomical units to express the distance from the sun to Pluto. At one point in Pluto's orbit, it is 4,000,000,000 miles from the sun. All I have to do is divide by the distance in an astronomical unit!

4,000,000,000 ÷ 92,955,807 = 43.03 astronomical units

That wasn't too complicated. Now if someone asked about the distance from the sun to Pluto, I can say it is about 43 astronomical units or about 43 distances from the Earth to the sun.

That doesn't help me all that much because the distance to Alpha Centauri is over 250,000 astronomical units. I hope there is a better way to express REALLY long distances.

You are in luck! The light-year is a unit of measure that is made for very, very long distances like the distance to stars. A light-year is the distance light travels in one year.

If you remember, the speed of light is approximately 186,000 miles per second. With that kind of speed, you can imagine how far light would travel in one year!

186,000 miles x 60 seconds per minute
x 60 minutes in an hour
x 24 hours in a day
x 365 days in a year

When I multiply that out, I get about 6 trillion miles. Light can travel pretty far in one year! Now I can express the distance to Alpha Centauri in light-years instead of miles or astronomical units.

250,000 astronomical units x
92,955,807 miles per astronomical unit =
23,238,951,750,000 miles to Alpha Centauri ÷
6 trillion miles per light-year = Approximately 4 light-years

Try the following problems:

1) When Earth and Saturn are closest, they are approximately 750,000,000 miles apart. How many astronomical units is Earth from Saturn? (Round to the nearest whole number.)

2) How many astronomical units are in one light-year? (Round to the nearest five thousand.)

3) A light-year is approximately 6 trillion miles. What is the distance of a light-hour? (Round to the nearest million.)

4) The distance to the moon is approximately 250,000 miles. What fraction of an astronomical unit is this?

   a) $\frac{1}{1,255,000}$ b) $\frac{1}{572,000}$ c) $\frac{1}{372}$ d) $\frac{1}{82}$

3 teaspoons = 1 tablespoon
2 tablespoons = 1 fluid ounce
8 fluid ounces = 1 cup
2 cups = 1 pint
2 pints = 1 quart
4 quarts = 1 gallon

12 inches = 1 foot
3 feet = 1 yard
5280 feet = 1 mile
92,955,807 miles = one astronomical unit
63,241 astronomical units = one light-year

16 ounces = 1 pound
2000 pounds = 1 ton

Problem Set 1

Warmup: If a 4 foot deep bathtub is drained at a rate of one foot every 15 seconds, how long until the tub is drained?

Level 1: If a bathtub is filled at a rate of 3 inches per minute, how many seconds until it is filled one inch deep?

Level 2: If a bathtub is drained at a rate of 1½ gallons per minute, how many quarts are drained in 10 seconds?

Level 3: If an 18 inch deep bathtub is filled at the rate of 1½ inches per minute, what fraction of the tub will be filled in 8 minutes?

Genius Level: A hose is filling a 70 gallon tub at the rate of 2 gallons per minute. At the same time, an open drain is draining the tub at a rate of one quart per minute. How long until the tub is filled?

Problem Set 2

Warmup: A recipe for one dozen cookies calls for 2 eggs. If you want to make 48 cookies, how many eggs would you need?

Level 1: A recipe for 10 cookies calls for 2 eggs. If you want to make 25 cookies, how many eggs would you need?

Level 2: If a recipe for 2½ pounds of cookies calls for ¼ teaspoon of baking powder, how much baking powder would you use for 1¼ pounds of cookies?

Level 3: If a recipe for 48 cookies calls for 1 cup of olive oil, how many teaspoons of olive oil would you need for 17 cookies?

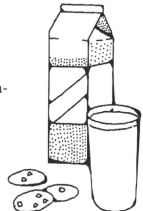

Genius Level: Holly was making a cake and put 5 cups of oats in a mixing bowl instead of the 4 cups that the recipe called for. Now she has to change all the other amounts in the recipe. What are the new amounts?

Recipe		New recipe
Oats	4 cups	5 cups
Honey	2 tablespoons	?
Oil	½ cup	?
Milk	6 fluid ounces	?

Problem Set 3

Warmup: If a snail crawls 2 feet per hour, how long does it take for the snail to crawl 2 yards?

Level 1: If a snail crawls 6 inches in one hour, how many yards will it crawl in 24 hours?

Level 2: If a sloth moves 2 yards in 15 minutes, how many days will it take the sloth to move one mile? (Round to the nearest whole day.)

Level 3: If a leopard runs at a speed of 60 miles per hour, how far will it run in one minute?

Genius Level: At a track meet, Jackie ran a mile in 6 minutes. What was her speed in miles per hour?

Problem Set 4

Warmup: If the Earth is 93,000,000 miles from the sun, how many astronomical units is the Earth from the sun?

a) 1 unit   b) 10 units   c) 9.3 units   d) 93 units

Level 1: How many inches are in one mile?

Level 2: Pluto is approximately 3,700,000,000 miles from the sun. How many astronomical units is Pluto from the sun? (Round to the nearest whole number.)

Level 3: A yard is what fraction of a mile?

Genius Level: If a spacecraft travels at 20,000 miles per hour, how many years will it take to travel one light-year? (The speed of light is 186,000 miles per second.)

a) 33 years   b) 334 years   c) 3348 years   d) 33,480 years

Problem Set 5

Warmup: How many cups are in one quart?

Level 1: How many teaspoons are in one fluid ounce?

Level 2: A recipe called for 2½ cups of sugar. The only measuring equipment Rachel could find was a tablespoon. How many tablespoons of sugar should Rachel use?

Level 3: Many chainsaws require a gasoline and oil mixture instead of just gasoline. Stacy bought a chainsaw that requires 1 ounce of oil for every 32 ounces of gasoline. Stacey needs to know how many fluid ounces of oil to put in one gallon of gas. (Be careful because if you give Stacey the wrong answer, it might ruin her new chainsaw.)

Genius Level: There are 4 quarts in one gallon and ¼ of a gallon is a quart. There are 2 pints in a quart and ½ quart in one pint. How many quarts are in one teaspoon?

Measurement
Level 1

1) If a ½ gallon milk jug leaks one cup an hour, how long until it is empty?

2) A recipe for 20 cookies calls for 6 cups of oats. How many cups of oats would be needed for 30 cookies?

3) Giant ants that weigh one pound are each carrying an egg that weighs 3 ½ pounds. If the weight limit of a bridge is 20 pounds, how many egg carrying ants can cross the bridge at the same time?

4) There are 4 quarts in one gallon. What fraction of a gallon is one quart?

5) How many inches are in ¼ mile?

Measurement
Level 2

1) A gallon of water will last 2½ days for 4 people. How long will it last for one person?

2) A recipe for 30 cupcakes calls for one quart of milk. If you needed to make 45 cupcakes, how many pints of milk would you need?

3) An ant is crawling along a yardstick at the rate of 1⅓ inches per second. How many seconds until the ant travels the length of the yardstick?

4) What fraction of a gallon is a cup?

5) A recipe for 20 cookies calls for ½ teaspoon of salt. If you are going to make 5 cookies, how much salt should you use?

Measurement
Level 3

1) Carolyn can walk 1½ miles in 60 minutes. How far can she walk in 70 minutes?

2) What fraction of a kilometer is a millimeter? *(Research needed)*

3) If milk cost $6 per gallon, what is the cost of each cup of milk?

4) If a sloth moves at a speed of 10 feet per hour and it travels 12 hours per day, how many days will it take to move one mile?

5) A skittish mouse is trying to cross a 100-yard football field. Each hour it travels 20 yards and then turns around and goes back 10 yards. How many hours will it take for the mouse to reach the end of the 100-yard field?

Measurement
Genius Level

1) Samantha is mixing ingredients to make deer repellent. The instructions say to mix 4 fluid ounces of "No Deer" with one gallon of water. Samantha only wants to make one cup of deer repellent. How many teaspoons of "No Deer" should Samantha use with one cup of water?

2) If Kath can walk 1½ miles in 60 minutes, how far can she walk in 75 minutes?

3) Each strand of Rapunzel's hair can hold ⅔ of an ounce. If Rapunzel had 3600 strands of hair, can a 175 pound witch safely climb up Rapunzel's hair? (Give the witch an engineering report to advise her.)

4) What fraction of a gallon is a teaspoon?

5) A hose is filling a 6 gallon jug at the rate of one quart per minute. At the same time, a hole in the jug is causing the jug to lose water at the rate of one cup per minute. How many minutes until the 6 gallon jug is filled with water?

Chapter Fifteen
Perimeter & Circumference

Finding the distance around things is a very important area of math. Let's say you were going to plant a garden and grow green beans. Rabbits love to eat green bean plants so you would need to build a fence. If your garden was this shape, it would be very easy to figure out how much fence you would need. All you need to do is add all 4 sides = 130 feet.

15 feet

50 feet   **50 feet**

15 feet

What if your garden was this shape? Now it is a little harder to determine how much fence you would need. You need to be a little bit clever to find the lengths of the missing sides.

8 feet

6 feet

10 feet   **???**

?

20 feet

You don't have to be that clever to find the missing sides. All I did was subtract 8 feet from 20 feet to find the length of the ??? side. Then I subtracted 6 feet from 10 feet to find the ? side. Now it is easy to find the perimeter: 10 + 8 + 6 + 12 + 4 + 20 = 60 feet.

8 feet

6 feet

10 feet   **12 feet**

4 feet

20 feet

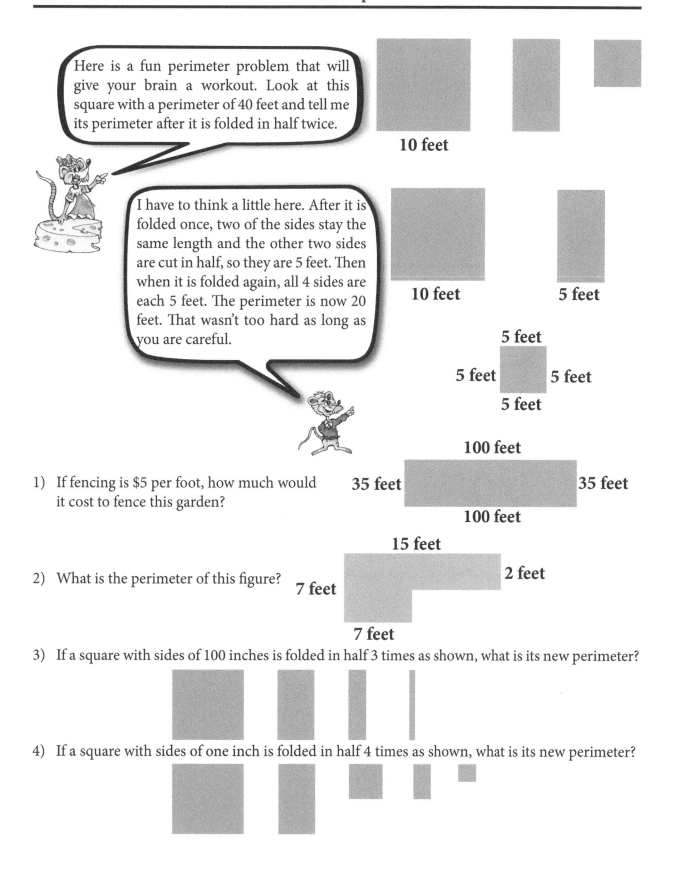

Here is a fun perimeter problem that will give your brain a workout. Look at this square with a perimeter of 40 feet and tell me its perimeter after it is folded in half twice.

10 feet

I have to think a little here. After it is folded once, two of the sides stay the same length and the other two sides are cut in half, so they are 5 feet. Then when it is folded again, all 4 sides are each 5 feet. The perimeter is now 20 feet. That wasn't too hard as long as you are careful.

10 feet   **5 feet**

5 feet
5 feet   **5 feet**
5 feet

100 feet

1) If fencing is $5 per foot, how much would it cost to fence this garden?

35 feet   **35 feet**

100 feet

15 feet

2) What is the perimeter of this figure?   **2 feet**

7 feet

7 feet

3) If a square with sides of 100 inches is folded in half 3 times as shown, what is its new perimeter?

4) If a square with sides of one inch is folded in half 4 times as shown, what is its new perimeter?

When we find the distance around rectangles and squares, we are finding the perimeter. When we are talking about the distance around a circle, there is a special name for it. It is called the circumference.

Finding the circumference of a circle is very easy. All you have to do is multiply the diameter of the circle by pi. Pi is a very special number. So special that it has its own symbol: π Oh, and by the way, if you do not know what a diameter is, it is the distance across the middle of a circle.

π is equal to 3.14159265359.......
The digits in pi go on and on forever so we usually round it off to 3.14. Look how easy it is to find the circumference of this circle.

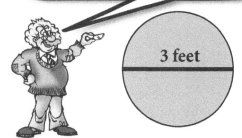

3 feet

3.14 x 3 feet = 9.42 feet

1) What is the distance around a circular table if the distance across the table is 4 feet?

2) If a snail can crawl at a speed of ½ foot per hour, how long will it take it to crawl all the way around a circular lake that is 100 feet across?

100 feet

3) If a circle with a diameter of 10 feet is folded in half twice as shown, what is the perimeter of the resulting figure?

4) How far does the point of a 7 inch minute hand travel from 9:00 to 9:45?

Pythagorean Triples

A Pythagorean triple triangle is a right triangle with special measurements that make it very easy to find the lengths of all three sides of the triangle. In case you have forgotten, a right triangle is one with one 90 degree angle which is marked with the symbol shown on this triangle.

I've placed some of the most common Pythagorean triples on the blackboard. I will show you how to use them after you take a look at them.

```
3 : 4 : 5
5 : 12 : 13
8 : 15 : 17
7 : 24 : 25
9 : 40 : 41
```

Look at this Pythagorean triple triangle. I know it is a Pythagorean triple because it is a right triangle and 2 of its sides match one of the Pythagorean triples. Now I know the length of the missing side without even measuring. It must equal 5 inches.

4 inches   **? inches**

3 inches

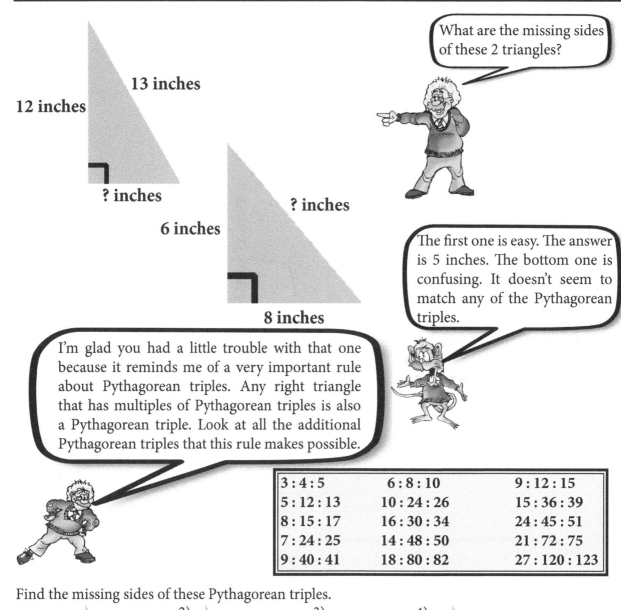

13 inches

12 inches

? inches

6 inches

? inches

8 inches

What are the missing sides of these 2 triangles?

The first one is easy. The answer is 5 inches. The bottom one is confusing. It doesn't seem to match any of the Pythagorean triples.

I'm glad you had a little trouble with that one because it reminds me of a very important rule about Pythagorean triples. Any right triangle that has multiples of Pythagorean triples is also a Pythagorean triple. Look at all the additional Pythagorean triples that this rule makes possible.

3 : 4 : 5	6 : 8 : 10	9 : 12 : 15
5 : 12 : 13	10 : 24 : 26	15 : 36 : 39
8 : 15 : 17	16 : 30 : 34	24 : 45 : 51
7 : 24 : 25	14 : 48 : 50	21 : 72 : 75
9 : 40 : 41	18 : 80 : 82	27 : 120 : 123

Find the missing sides of these Pythagorean triples.

1)   36   39   ?

2)   9   ?   12

3)   90   ?   120

4)   40   ?   9

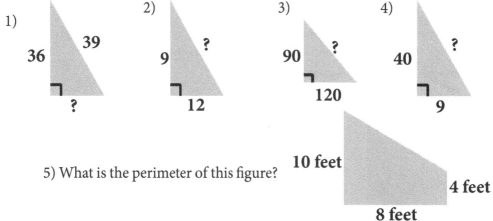

5) What is the perimeter of this figure?

10 feet

4 feet

8 feet

Problem Set 1

Warmup: What is the perimeter of this rectangle?

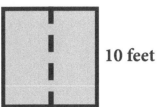

15 feet

5 feet

Level 1: The square shown is folded in half at the dotted line. What is the perimeter of the new rectangle?

10 feet

Level 2: The rectangle shown below is folded in half twice. What is the perimeter of the new rectangle?

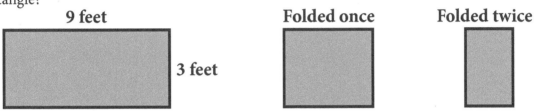

9 feet

3 feet

Folded once

Folded twice

Level 3: The rectangle shown below is cut along the diagonal as shown. What is the perimeter of each new triangle?

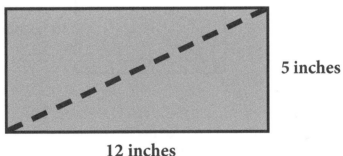

5 inches

12 inches

Genius Level: A circle with a 10 inch diameter is folded in half twice. What is the perimeter of the resulting figure?

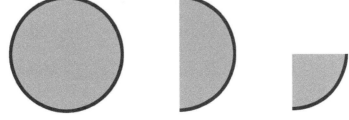

Problem Set 2

Warmup: A worm crawls around the outside of a square that is 15 feet on each side. If its speed is one foot per minute, how many hours will it take the worm to crawl around the entire square?

15 feet

Level 1: A worm crawls around the outside of a rectangle that is 15 feet wide and 30 feet long. If its speed is one foot per minute, how many hours will it take the worm to crawl around the entire rectangle?

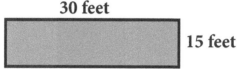

30 feet

15 feet

Level 2: The length of a rectangle is three times its width. The perimeter of the rectangle is 40 inches. What is the width of the rectangle?

Level 3: Wally the worm travels from the bottom left corner of a 30 inch by 40 inch rectangle to the top right corner at a speed of ½ inch per minute. How long does it take for Wally's trip?

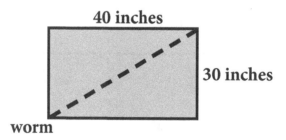

40 inches

30 inches

worm

Genius Level: Two worms are resting in the lower right corner of a 20 inch by 80 inch rectangle. Worm A starts traveling up the right side of the rectangle at a speed of 3 inches per minute and worm B starts traveling the opposite direction along the rectangle at a speed of 2 inches per minute. How long until they meet?

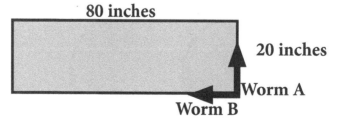

80 inches

20 inches

Worm A

Worm B

Problem Set 3

Warmup: What is the perimeter of this figure?

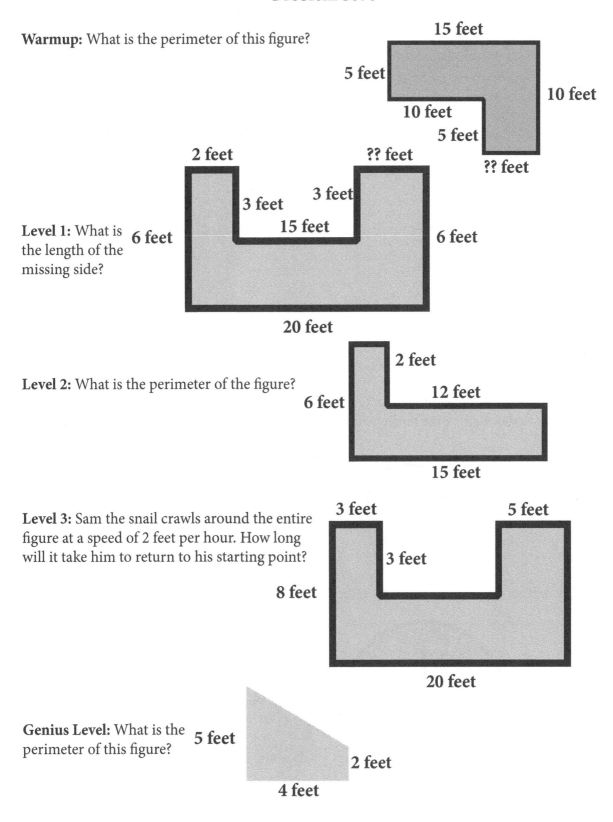

Level 1: What is the length of the missing side?

Level 2: What is the perimeter of the figure?

Level 3: Sam the snail crawls around the entire figure at a speed of 2 feet per hour. How long will it take him to return to his starting point?

Genius Level: What is the perimeter of this figure?

Problem Set 4

Warmup: If the diameter of a circle is 100 inches, what is the circumference?

Level 1: If the radius of a circle is 5 inches, what is the circumference?

Level 2: The ratio of the circumference of a circle to its diameter is equal to _____.

   Circumference/Diameter = ?

Level 3: Andy the ant walks around a circle with a diameter of 30 feet while Aubrey the ant walks around a circle with a diameter of 60 feet. How many feet longer is Aubrey's trip than Andy's?

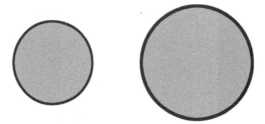

Genius Level: How far does the point of a 6 inch minute hand travel from 9:00 to 9:20?

Problem Set 5

Warmup: The Earth's circumference is approximately_____times its diameter.

      a) 2   b) 3   c) 4   d) 5   e) Impossible to tell

Level 1: The moon's diameter is approximately 2160 miles. Estimate the moon's circumference by using the formula $\pi \times D = C$. (Round to the nearest hundred.)

Level 2: Kristin looked at a round table in her cafeteria and said that she could predict the result if you measured the distance around the table and divided it by the distance across the table. Distance around the table/distance across the table = ?

What number did Kristin predict?

Level 3: How far does the tip of a 6 inch long hour hand travel in an 18 hour period?

Genius Level: Jacob wants to see if it is faster to canoe across a round lake or walk around the outside of the lake. He found that it is 3.14 miles from Point A to Point B when he walks around the lake. Jacob walks at 3 miles per hour, so he will walk the 3.14 miles to the other side of the lake in a little over one hour. How long will it take him to canoe from one side of the lake to the other side if he canoes at 2 miles per hour?

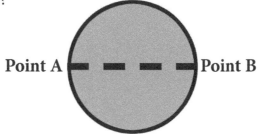

Point A   Point B

Perimeter & Circumference
Level 1

1) A rectangle has a 26 inch perimeter and a 3 inch width. What is the length of the rectangle?

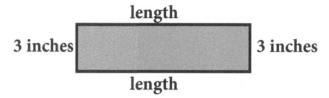

2) What is the length of each side of a square that has a perimeter of 84 inches?

3) What are the lengths of sides A and B?

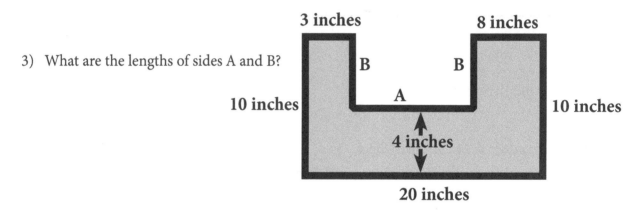

4) The 8 inch square shown below is folded along the dotted line and then folded again along the dotted line. What is the perimeter of the new square?

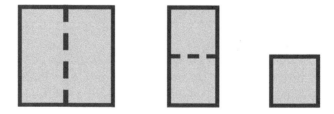

5) Some large redwood trees have trunks that are 20 feet across. What is the distance around a tree that is 20 feet across?

Perimeter & Circumference
Level 2

1)   Maria measured the distance around a circular table and found that it was 9.42 feet. She said that with her magic powers she could tell the distance across the table without measuring. What is the distance across the table?

2)   A duck needs to get to the other side of a circular lake. Because it is worried about bass eating it if it swims across the lake, the duck decided to walk around the edge of the lake to get to the other side. If the lake is 400 feet across, how many feet longer will the duck's trip be because it decided not to swim across the lake?

3)   This rectangle is folded in half along the dotted line. What is the perimeter of the new rectangle?

3 inches

8 inches   **8 inches**

3 inches

4)   What is the perimeter of the figure shown?

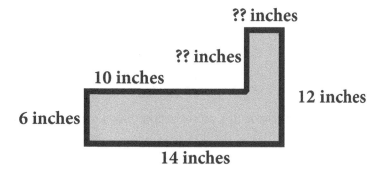

?? inches

?? inches

10 inches

12 inches

6 inches

14 inches

5)   If a snail travels at a speed of 3 inches per minute, how long will it take for it to travel across a circular table with a circumference of 37.68 inches?

Perimeter & Circumference
Level 3

1) Marlene wants to build a circular table with enough room for 6 people. If each person requires 3.14 feet of space, what diameter should Marlene make the table?

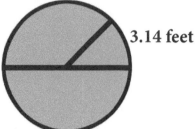

3.14 feet

2) A circle with a diameter of 10 inches is folded in half along the dotted line. What is its new perimeter?

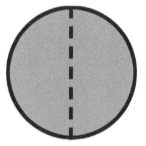

3) If a superworm crawls at a speed of 1.5 inches per second, how long will it take the worm to crawl around the outside of a table with a 50 inch diameter?

50 inches

4) What is the perimeter of this figure?

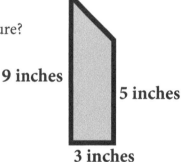

9 inches

5 inches

3 inches

5) A duck needs to get to the other side of a circular lake. Because it is worried about bass eating it if it swims across the lake, the duck is thinking about walking around the lake. This is what we know:

a) The lake is 400 feet across
b) The duck can swim at a speed of 20 feet per minute
C) The duck can walk at a speed of 30 feet per minute

What choice will get the duck to the other side faster, swimming across or walking around the edge? Why?

Perimeter & Circumference
Genius Level

1) A coyote walked around the outside of a circular lake at a speed of 5 miles per hour. If it took the coyote one hour and 12 minutes to complete the trip, what is the distance across the lake? (Round to the nearest tenth of a mile.)

2) A circle with a diameter of 100 inches is folded in half along the dotted line, folded in half along the new dotted line, and then folded one more time. What is the perimeter of the new figure?

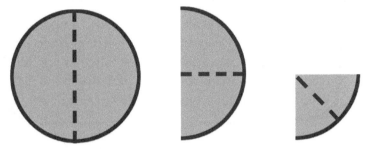

3) How long will it take a sloth that is running at a speed of one foot per minute to run once around this track? (The ends of the running track are semicircles.)

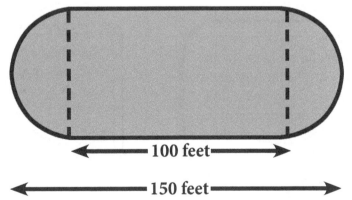

4) How far does the tip of a 5 inch long minute hand travel in a 10 hour period?

5) The tip of an hour hand on a clock travels 37.68 inches in a 24 hour period. How long is the hour hand of the clock?

Chapter Sixteen
Area

Finding areas is a very important skill that students must learn. It helps in many areas of everyday life such as buying carpet and painting walls. It is also a very important skill that is required in many jobs such as engineering, architecture, construction, many medical fields, retail, education, and hundreds more.

I know another reason understanding areas is important. To pass math tests!

You're right that learning about areas is important for passing math tests, but the reason it is important to pass math tests is that they test whether you are ready to do the math you will experience in the real world. Sorry, I think I just sounded like an adult.

Let's start with finding areas of rectangles and squares. Look at the problem on the board. Gabriel has to determine how many tiles he needs to buy, but he doesn't want to do a lot of counting. Finding the area of the room will help him determine the number of tiles to buy very quickly.

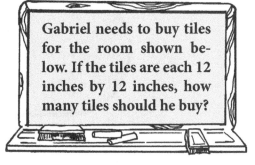

Gabriel needs to buy tiles for the room shown below. If the tiles are each 12 inches by 12 inches, how many tiles should he buy?

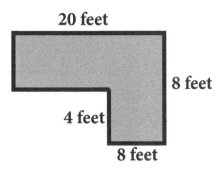

20 feet

8 feet

4 feet

8 feet

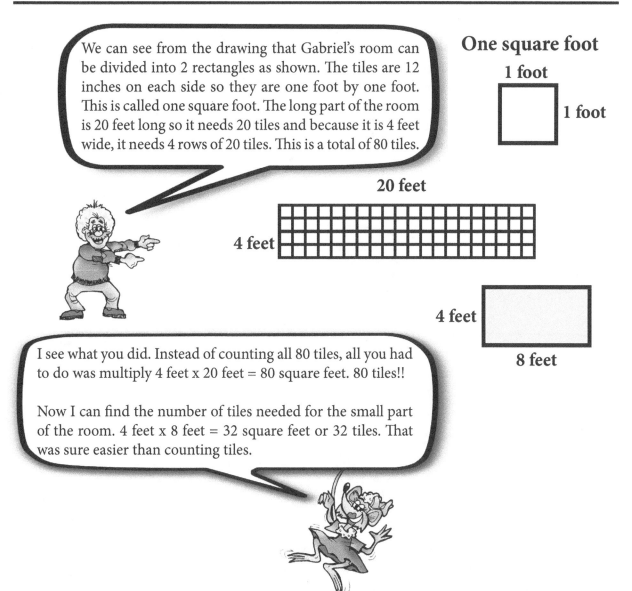

We can see from the drawing that Gabriel's room can be divided into 2 rectangles as shown. The tiles are 12 inches on each side so they are one foot by one foot. This is called one square foot. The long part of the room is 20 feet long so it needs 20 tiles and because it is 4 feet wide, it needs 4 rows of 20 tiles. This is a total of 80 tiles.

One square foot

1 foot

1 foot

20 feet

4 feet

I see what you did. Instead of counting all 80 tiles, all you had to do was multiply 4 feet x 20 feet = 80 square feet. 80 tiles!!

Now I can find the number of tiles needed for the small part of the room. 4 feet x 8 feet = 32 square feet or 32 tiles. That was sure easier than counting tiles.

4 feet

8 feet

1) What is the area of a room that is 25 feet long and 10 feet wide?

2) If carpet is $12 per square foot, what would it cost to carpet a room that is 12 feet long and 12 feet wide?

3) What is the area of this room?

40 feet

18 feet

10 feet

15 feet

Areas of Circles

The area of a circle is as easy to find as the area of rectangles. Before you can find the areas of circles you must learn what the radius of a circle is.

The radius of a circle is the length from the center to the outside of the circle. It is half of the diameter of the circle.

To find the area of a circle, all you have to do is use this formula. Remember that pi is equal to 3.14. So the area of a circle with a radius of 5 inches is: 3.14 x 5 x 5 = 78.5 square inches.

Pi x radius x radius = area

There is something you need to know about circles to help you find their areas. Circles are made up of 360 degrees. Look at these three circular floors that need to be painted. Each has a radius of 8 feet.

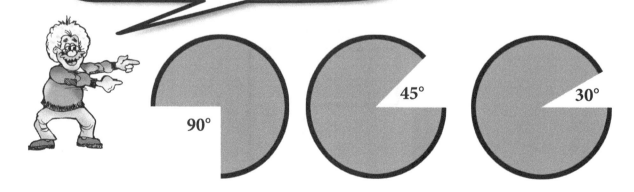

90°   45°   30°

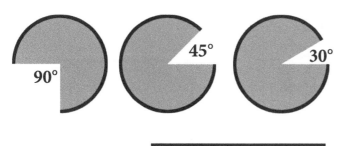

I see that each floor has only a part of it to be painted. The first one has ¾ of the circle to be painted because 90° is ¼ of the circle. So I find the area of the circle and then take away ¼ of that area. 150.72 square feet need to be painted.

3.14 x 8 x 8 = 200.96

200.96 ÷ 4 = 50.24

200.96 - 50.24 = 150.72

The second floor is a little more confusing. It has 45° that will not be painted. That is ⁴⁵/₃₆₀ or ⅛ of the floor to subtract from the area. 175.84 square feet need to be painted.

3.14 x 8 x 8 = 200.96

200.96 ÷ 8 = 25.12

200.96 - 25.12 = 175.84

I get the idea. The 3rd circle has ³⁰/₃₆₀ = ¹/₁₂ that will not be painted. 184.21 square feet need to be painted.

3.14 x 8 x 8 = 200.96

200.96 ÷ 12 = 16.75

200.96 - 16.75 = 184.21

1) What is the area of a circle with a radius of 20 feet?

2) If the cost of paint is $40 per 100 square feet, what would it cost to paint a circular floor with a radius of 10 feet?

3) A circle with a radius of 60 feet needs to be painted. If 120° of the circle will not be painted, how many square feet will be painted?

4) This square has sides that are 10 feet long. The circle is to be painted light green and the rest is to be painted dark green. How many square feet will be painted dark green?

Areas of Triangles

The area of this rectangle is very easy to find. 10 feet x 5 feet = 50 square feet. Now, with a little trick, it is easy to find the area of a triangle. What do you think the area of the dark triangle is equal to?

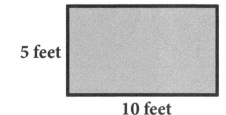

5 feet

10 feet

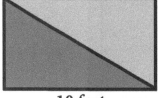

5 feet

10 feet

That is so easy, I can't believe I didn't know how to find areas of triangles! All I do is take the area of the rectangle and divide it in half. The area of this triangle is 25 square feet.

To find the area of any triangle, just find the height and then multiply by the base and then divide by two. In our example, the height was very easy to find. There are a couple other types of triangles whose height is a little more difficult to find such as the two triangles I have drawn here. The height is the dotted line and it is perpendicular to the base.

Find the area of these figures:

1)

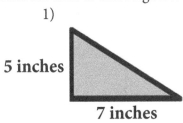

5 inches

7 inches

2)

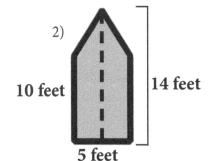

10 feet

14 feet

5 feet

3)

5 inches

4 inches

Problem Set 1

Warmup: What is the area of a rectangle that is 3 feet wide and 7 feet long?

Level 1: What is the area of a circle with a radius of 5 feet?

Level 2: What is the area of the figure shown below?

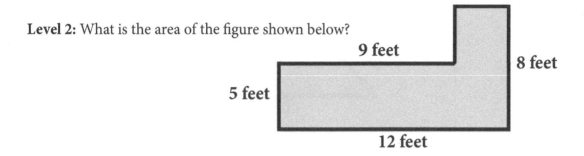

Level 3: What is the area of the figure below?

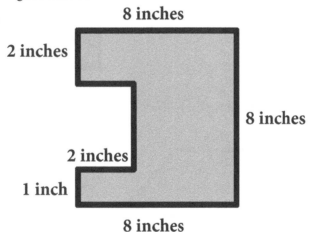

Genius Level: What is the area of the shaded part of the figure below if the diameter of the circle is 100 feet?

Problem Set 2

Warmup: What is the area of the triangle shown?

5 inches

10 inches

Level 1: What is the area of the triangle shown below?

8 inches

20 inches

Level 2: What is the area of the figure shown below?

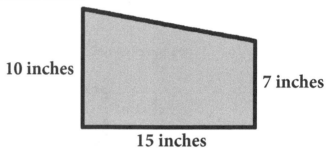

10 inches

7 inches

15 inches

Level 3: What is the area of the figure below?

10 inches

6 inches

14 inches

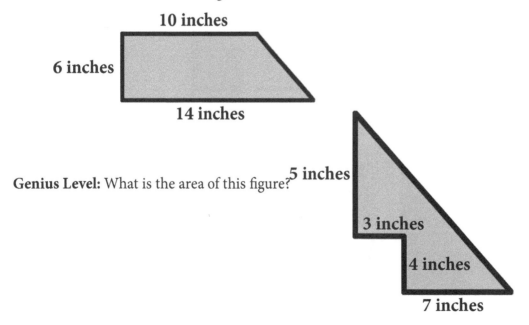

Genius Level: What is the area of this figure? 5 inches

3 inches

4 inches

7 inches

Problem Set 3

Warmup: If carpet is $10.50 per square yard, what would it cost to carpet a room that has an area of 25 square yards?

Level 1: If carpet is $3 per square foot, what would it cost to carpet a room that is 12 feet wide and 15 feet long?

Level 2: How many 12 inch x 12 inch tiles would you need to buy for a room that is 18 feet wide and 24 feet long?

Level 3: A dog is tied in the center of a carpeted 25 foot x 25 foot room on a 10 foot chain. The dog has ruined the carpet everywhere it can reach. If carpet is $5 per square foot, what will it cost to replace the part of the carpet that is ruined?

Genius Level: If carpet is $32 per square yard, what would it cost to carpet the room shown below?

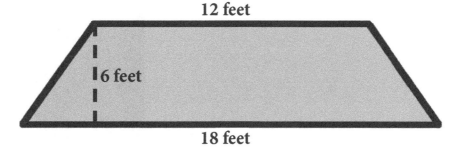

12 feet

6 feet

18 feet

Problem Set 4

Warmup: Jacob needed to paint a floor that was 15 feet long and 15 feet wide. The paint he was using could cover 125 square feet for each gallon. How many gallons of paint should Jacob buy?

a) 1   b) 2   c) 3   d) 4

Level 1: Samantha is painting a fence that is 20 feet long and 8½ feet high. The paint she is using will cover 85 square feet per gallon. How many gallons should she buy?

Level 2: There is a wall that needs to be painted. It is 36 feet long and 8 feet tall. The wall has a window in it that is 7 feet tall and 9 feet wide that does not need to be painted. If the paint covers 75 square feet per gallon, how many gallons are needed?

Level 3: This wall needs to be painted. If the paint covers 56 square feet per gallon, how many gallons of paint are needed?

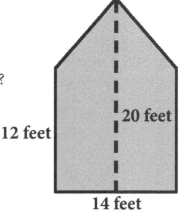

Genius Level: The outside of a cylindrical silo needs to be painted. The silo is 80 feet tall and 30 feet across. If the paint covers 161 square feet per gallon, how many gallons are needed to paint the outside of the silo?   (The top and bottom will not be painted.)

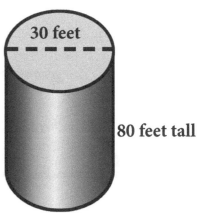

Problem Set 5

Warmup: What is the area of a circle with a radius of 5 inches?

Level 1: What is the area of the missing part of the circle shown below if the radius is 5 inches?

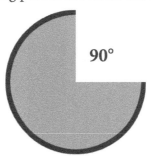

Level 2: What is the area of the missing part of the circle shown if the radius is 5 inches? (Hint: There are 360 degrees in a circle.)

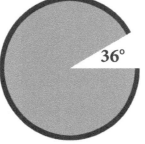

Level 3: What is the area of the missing parts of the circle if the radius is 5 inches?

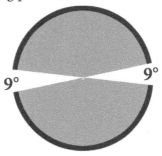

Genius Level: What is the area of the missing part of the circle? The radius of the circle is 5 inches.

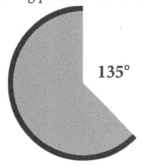

Areas
Level 1

1) A rug that is 8 feet wide and 15 feet long cost $360. How much is that per square foot?

2) How many square feet are in 5 square yards?

3) How many square feet of paint do you need for a circular room that is 20 feet across?

4) How many 4 inch by 4 inch tiles are needed to tile a table top that is one square foot?

5) The area of a rectangular floor is 286 square feet. If the length of the floor is 22 feet, what is the width?

Areas
Level 2

1) If carpet is $30 per square yard, what does it cost to carpet a room that is 18 feet wide and 36 feet long?

2) How many tiles that are 12 inches by 18 inches will be needed to tile a floor that is 15 feet wide and 20 feet long?

3) How many square feet are in the room shown below?

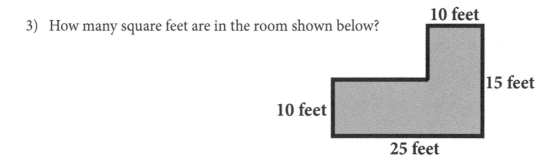

4) How many stamps that are 1.5 inches long and 1.5 inches wide will fit on a piece of paper that is 9 inches wide and one foot long?

5) If it cost $1,080 to carpet a room that is 12 feet by 18 feet, what was the cost of the carpet per square yard?

Areas
Level 3

1) What is the area of the shaded part?

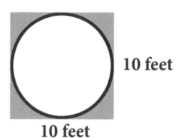

10 feet

10 feet

2) If carpet is $41.40 per square yard, what would it cost to carpet a room that is 20 feet wide and 30 feet long?

3) If you wanted to tile all 6 sides of a cube that is one yard long on each side, how many square feet of tiles would be needed?

4) A certain kind of dice are one inch tall, one inch wide and one inch long. How many of the dice can be fit on the floor of a room that is 10 feet wide and 12 feet long?

5) If paint is $17.50 per quart and each quart covers 38 square feet, what is the cost of the paint to cover the wall shown below?

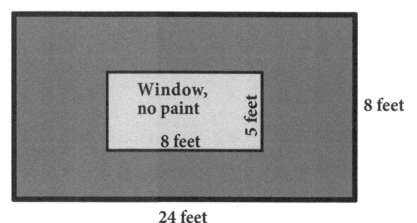

Window, no paint

8 feet

5 feet

8 feet

24 feet

Areas
Genius Level

1) What is the area of this figure?

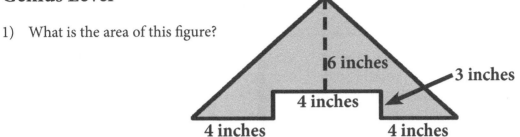

2) If 4 inch by 4 inch tiles cost 89 cents each, what is the cost to tile the entire outside of a cube that is one yard on each side?

3) A garden sprinkler is broken and only waters the lightly shaded semicircle section of a square garden. Bria wants to put decorative rocks in the part of the garden that doesn't get watered. If the rocks cost $1.20 per square foot, what is the total cost of the rocks?

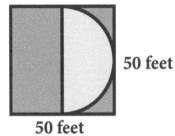

4) What is the area of the floor shown below?

18 feet

12 feet

36 feet

5) The inside of a circular room will be painted a light color and the outside will be painted brown. The room is 20 feet in diameter and the inside part is 10 feet in diameter. Determine how many square feet of brown paint you will need.

Chapter Seventeen
Volume

All objects have volume because they take up space. Look at the problem on the board and then look at the dice that are one inch long, one inch wide and one inch tall.

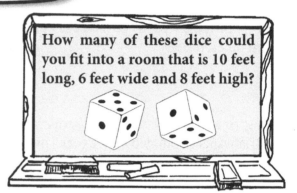

How many of these dice could you fit into a room that is 10 feet long, 6 feet wide and 8 feet high?

I used to do these problems for fun. The volume of each of the dice is one cubic inch. To find the volume of the dice, I multiplied length x width x height. 1 x 1 x 1 = 1 cubic inch. This means that each die takes up one cubic inch of space.

To solve the problem, you could buy thousands of dice and count how many would fit in the room, or you could use math to find the volume of the dice and the room and find the answer in one minute.

Now all I have to do is find the volume of the room. I first need to change the measurements of the room into inches so when I multiply to find the volume, my answer will be in cubic inches.

10 feet = 120 inches
8 feet = 96 inches
6 feet = 72 inches
Volume = 120 x 96 x 72 = 829,440 cubic inches
Dice in Room = 829,440

I see what you did. The room is 829,440 cubic inches and each of the dice takes up one cubic inch. Now I know that I can fit 829,440 dice into the room. Look how easily I found how many dice that are 2 inches on each side will fit into the room.

Volume of the room: = 829,440 cubic inches
Volume of each die: 2 x 2 x 2 = 8 cubic inches
Dice in room: 829,440 ÷ 8 = 103,680

1) How many cubic inches are in a giant die that is 6 inches on each side?

2) How many wood blocks that are 2 inches on each side will fit into a box that is a yard long on each side?

3) How many cubic decimeters are in a large box that is 5 meters long, 3 meters wide and 2 meters tall? (Remember that there are 10 decimeters in one meter.)

4) How many one decimeter cubes can fit inside a box that is one meter on each side? (The cubes are one decimeter on each side.)

Here is an interesting fact about the metric system. A liter of water is one cubic decimeter of water. So if I have a cube that is 5 decimeters on each side, its volume is 5 x 5 x 5 = 125 cubic decimeters. I also know it is 125 liters.

Here is an additional interesting fact about the metric system. A liter of water weighs one kilogram. Now we know that one cubic decimeter = one liter = one kilogram.

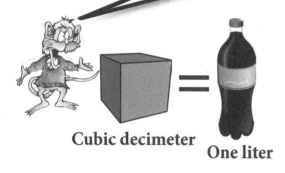

One liter

Cubic decimeter

One liter

One kilogram

Those connections make the metric system easier to work with than our system sometimes. If I have 10 cubic decimeters of water, I have 10 liters and I know it weighs 10 kilograms.

I once found the volume of a large metal sculpture. A friend said it was impossible to find the volume of the sculpture because it had a weird shape. I slowly dropped it into a full pool of water and then weighed the water that overflowed the pool. It weighed 19 kilograms so I knew that the volume of the sculpture was 19 cubic decimeters.

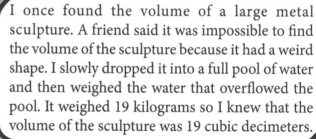

1) If a cubic decimeter metal block is dropped into a full pool of water, what is the weight of the water that will overflow?

2) How many liters of water are in a rectangular swimming pool that is 50 decimeters long, 20 decimeters wide and 10 decimeters deep?

3) A gram equals ¹⁄₁₀₀₀ of a kilogram. What volume of water weighs one gram? (*Research needed*)

    a) Cubic meter   b) Cubic decimeter   c) Cubic centimeter   d) Cubic millimeter

4) How many liters are in a cubic meter of water and what does it weigh?

Volumes of Cylinders

There is one last type of volume problem I want to talk about. This is the volume of a cylinder such as a can. Look at this can and notice that it is easy to find the area of the circular top of the can by using the formula from the chapter on areas:

π x radius x radius

Radius: 2 inches
Height: 6 inches

Because this can has a radius of 2 inches, the area of the circular top of the can is:

3.14 x 2 x 2 = 12.56 square inches

Now to find the volume, all I have to do is multiply by the height of the can:

12.56 x 6 inches = 75.36 cubic inches

π x radius x radius x height

I always wondered what the volume of the water in my garden hose was. I think I can figure it out now. I'll start by writing the formula for volume of a cylinder.

The garden hose I have has a diameter on the inside of one inch and a length of 50 feet. When you think about it, the hose is really a very long cylinder, so I can find the volume of water in the hose by using the formula for volume of a cylinder.

π x radius x radius x height
Diameter: one inch
Radius: ½ inch
Height: 50 feet x 12 = 600 inches
Volume: 3.14 x ½ x ½ x 600 = 471 cubic inches

I had no idea a hose could hold that much water! 471 cubic inches is a lot more volume than the can.

1) What is the volume of a can that has a radius of 6 inches and a height of 8 inches?

2) What is the volume of a 10 inch high juice can with a 5 inch diameter?

3) What is the volume of a 100 foot long hose with an inside diameter of 2 inches?

4) If a cylinder has a volume of 785 cubic inches and the area of the top is 78.5 square inches, what is its height?

Problem Set 1

Warmup: Dice are small cubes. Their volume is closest in size to:

a) a cubic millimeter   b) a cubic decimeter   c) a cubic inch   d) a cubic foot

Level 1: How many cubic feet are in a cubic yard?

Level 2: How many cubic inches are in a cubic foot?

Level 3: How many cubic inches are in a cubic yard?

Genius Level: How many 1 foot by 1 foot by 1 foot blocks would fit inside a cubic mile space?

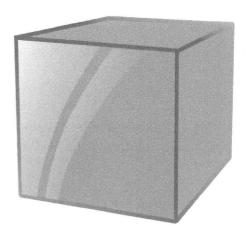

Problem Set 2

Warmup: What is the volume of a block that is 2 feet long, 2 feet wide and 2 feet tall?

Level 1: How many blocks that are 1 foot long, 1 foot wide and 1 foot tall will fit into a box that is 10 feet long, 3 feet wide and 3 feet tall?

Level 2: If sand is $12.50 per cubic yard, what is the cost of enough sand to fill a container that is 9 feet long, 3 feet wide and 3 feet tall?

Level 3: If the weight of the water in a hot tub that is 8 feet long, 54 inches wide and 42 inches deep is 7862.4 pounds, what does water weigh per cubic foot?

Genius Level: If cement is $75 per cubic yard, what does it cost to make a cement driveway that is 81 feet long, 15 feet wide and 9 inches thick?

Problem Set 3

Warmup: How many cubic decimeters are in this cubic meter?

Level 1: How many cubic centimeters are in a cubic meter?

Level 2: If a metal block 5 decimeters long by 3 decimeters wide by 2 decimeters tall was dropped into a full tank of water, how many liters of water would overflow?

Level 3: If a cubic meter of iron is dropped into a full tank of water that is 3 meters deep, what is the weight of the water that would overflow? (Give your answer in kilograms.)

Genius Level: A metal block that measures 80 centimeters by 70 centimeters by 60 centimeters is dropped into a full 3 meter deep tank of water. What is the weight of the water that overflows?

Problem Set 4

Warmup: How many cubic decimeters of water does it take to fill a liter bottle?

Level 1: If a hose can fill a liter container in one minute, how long will it take to fill a small swimming pool that is 120 cubic decimeters?

Level 2: Water comes out of a hose at the rate of a cubic decimeter every minute. How many minutes until the hose has filled a cubic meter?

Level 3: Multnomah Falls is a very high waterfall in Oregon. The water going over the falls averages 4 cubic meters per second. How many liter bottles can the falls fill every second?

Genius Level: The volume of water that goes over Horseshoe Falls is approximately 2,250,000 liters per second. How many swimming pools that are 50 meters long, 25 meters wide and 2 meters deep will the falls fill in one minute?

Problem Set 5

Warmup: The area of the top of this cylinder is 50 square inches. What is the volume of the cylinder?

Level 1: What is the volume of a cylinder with a radius of 5 inches and a height of 30 inches?

Level 2: A cylinder has a volume of 1256 cubic inches and the area of the top is 12.56 square inches. How tall is the cylinder?

Level 3: If a cylinder that is 5 inches high and has a 2 inch diameter is slowly dropped into a full pan of water that is 20 inches deep, what is the volume of the water that overflows?

Genius Level: What is the volume of the water in a full 100 foot hose that has an inside diameter of 1 inch?

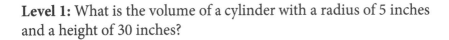

Volume
Level 1

1)   The volume of a cube is one cubic foot. What is the length of each side of the cube?

2)   How many cubic meters are in a block that is 2 meters long, 2 meters wide and 2 meters tall?

3)   What is the volume of a cement block that is 6 yards long, 2 yards wide and ½ yard tall?

4)   What is the volume of a 10 inch tall cylinder with a top that has an area of 10 square inches?

5)   A sink can hold 10 cubic feet of water. A faucet's flow is 3 cubic feet per minute and the drain empties 2 cubic feet per minute. How long until the sink is full?

Volume
Level 2

1) How many cubic feet are in 2 cubic yards?

2) If cement is $75 per cubic yard, what is the cost of a basement floor that is 30 feet by 45 feet and one foot thick? (Hint: What fraction of a yard is one foot?)

3) If a cylinder's volume is 1728 cubic inches, what is its volume in cubic feet?

4) The volume of a cube is 1000 cubic inches. What is the length of each side of the cube?

5) A waterfall has a flow rate of 50 cubic yards per second. How many hours does it take for 180,000 cubic yards of water to go over the waterfall?

Volume
Level 3

1) How many small cubes that are 1 centimeter on each side will fit into a room that is 10 meters long, 5 meters wide and 3 meters tall?

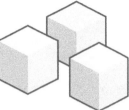

2) The volume of a cube is 1,000,000 cubic millimeters. How many cubic decimeters is the cube?

3) Water from a faucet fills a cubic foot box every 20 seconds. How many minutes will it take to fill a box that has a volume of one cubic yard?

4) What is the weight in kilograms of $\frac{1}{16}$ of a cubic decimeter of water?

5) A cubic foot container is full of water. A metal cylinder that is 10 inches tall with a 5 inch radius is placed on the bottom of the container. What is the volume of the water that will spill out of the cubic foot container?

Volume
Genius Level

1) The volume of a cube is ¹⁄₆₄ of a cubic yard. What is the length in inches of each side of the cube?

2) What is the volume of water in a 25 foot hose that has a diameter of ½ inch? (Round to the nearest cubic inch.)

3) What is the cost of a cement wall that is 7.5 feet high, 45 feet long and 9 inches thick if cement is $80 per cubic yard?

4) What is the weight, in kilograms, of the water in a pool that is 2 meters long, 2 meters wide and 2 meters deep?

5) If a waterfall's flow rate is 850,000 cubic feet per second, how long until a cubic mile goes over the falls?

    a) about 2 hours   b) about 2 days   c) about 2 months   d) about 2 years

Chapter Eighteen
Fun With Ratios

This chapter is meant to be fun. It is really just playing with fractions. Let's say I made a fraction with words:

$$\frac{\text{Number of noses a person has}}{\text{number of eyes a person has}} = \frac{?}{?}$$

All I have to do is fill in the numbers. The answer is ½. Look at these problems and see if you can come up with the answers.

$$\frac{\text{Cup}}{\text{Pint}} = \frac{?}{?}$$

$$\frac{\text{Pound}}{\text{Ton}} = \frac{?}{?}$$

$$\frac{\text{Meter}}{\text{Decimeter}} = \frac{?}{?}$$

To answer the first problem, I need to know how many cups are in a pint. I think there are two cups in a pint, so the answer is ½ because a pint is twice as big as a cup.

The next two are pretty simple because I know there are 2000 pounds in a ton and there are 10 decimeters in each meter.

$$\frac{\text{Pound}}{\text{Ton}} = \frac{1}{2000}$$

$$\frac{\text{Meter}}{\text{Decimeter}} = \frac{10}{1}$$

Sometimes I don't know what number goes in the numerator and what goes in the denominator. To keep from reversing them, I rewrite the problem like this.

$$\frac{\text{Pound}}{\text{2000 pounds (ton)}} = \frac{1}{2000}$$

$$\frac{\text{10 decimeters (meter)}}{\text{Decimeter}} = \frac{10}{1}$$

Fun with Ratios

Level 1

1) $\dfrac{\text{Number of legs on an ostrich}}{\text{Number of spider legs}} = \dfrac{?}{?}$

2) $\dfrac{\text{Foot}}{\text{Yard}} = \dfrac{?}{?}$

3) $\dfrac{\text{Inch}}{\text{Yard}} = \dfrac{?}{?}$

4) $\dfrac{\text{Quart}}{\text{Gallon}} = \dfrac{?}{?}$

5) $\dfrac{\text{Cup}}{\text{Fluid ounce}} = \dfrac{?}{?}$

6) $\dfrac{\text{Foot}}{\text{Mile}} = \dfrac{?}{?}$

7) $\dfrac{\text{Number of legs dog}}{\text{Number of legs grasshopper}} = \dfrac{?}{?}$

8) $\dfrac{\text{Ounce}}{\text{Pound}} = \dfrac{?}{?}$

9) $\dfrac{\text{Value of a quarter}}{\text{Value of a 20-dollar bill}} = \dfrac{?}{?}$

10) $\dfrac{\text{Value of a nickel}}{\text{Value of 3 quarters}} = \dfrac{?}{?}$

Fun with Ratios
Level 2

1) $\dfrac{12 \text{ minutes}}{\text{Hour}} = \dfrac{?}{?}$

2) $\dfrac{\text{Decimeter}}{\text{Meter}} = \dfrac{?}{?}$

3) $\dfrac{\text{Millimeter}}{\text{Decimeter}} = \dfrac{?}{?}$

4) $\dfrac{\text{Liter}}{\text{Milliliter}} = \dfrac{?}{?}$

5) $\dfrac{\text{Meter}}{\text{Kilometer}} = \dfrac{?}{?}$

6) $\dfrac{\text{Square foot}}{\text{Square yard}} = \dfrac{?}{?}$

7) $\dfrac{\text{Second}}{\text{Hour}} = \dfrac{?}{?}$

8) $\dfrac{\text{⅓ yard}}{1 \text{ inch}} = \dfrac{?}{?}$

9) (Number of legs dog/number of legs chicken) ÷ (Number of legs spider/number of legs fly) = ?

10) (Number of fingers on each hand/thumbs each hand) ÷ Number of legs on two spiders $= \dfrac{?}{?}$

Fun with Ratios

Level 3

1) $\dfrac{\text{Cup}}{\text{Gallon}} = \dfrac{?}{?}$

2) $\dfrac{\text{Cubic foot}}{\text{Cubic yard}} = \dfrac{?}{?}$

3) $\dfrac{\text{Weight of a cubic decimeter of water}}{\text{1 kilogram}} = \dfrac{?}{?}$

4) $\dfrac{\text{Kilometer}}{\text{Millimeter}} = ?$

5) $\dfrac{\text{Oxygen atoms in one molecule of water}}{\text{Hydrogen atoms in one molecule of water}} = \dfrac{?}{?}$

6) $\dfrac{\text{Number of degrees of all interior angles in a hexagon}}{\text{Number of degrees of all interior angles in a triangle}} = ?$

7) $\dfrac{\text{Oxygen atoms in one molecule of hydrogen peroxide}}{\text{Hydrogen atoms in one molecule of hydrogen peroxide}} = ?$

8) $\dfrac{\text{Gallon}}{\text{Tablespoon}} = ?$

9) (Ton/pound) ÷ (Pound/ounce) = **?**

10) $\dfrac{\text{Height in feet of the Empire State Building including antennas (round to the nearest 100)}}{\text{Height in feet of the Statue of Liberty including the pedestal (round to the nearest 100)}} = ?$
 (Research Needed)

Fun with Ratios
Genius Level

1) $\dfrac{\text{Circumference of circle A}}{\text{Diameter of circle A}} = ?$

2) $\dfrac{\text{Cubic millimeter}}{\text{Cubic meter}} = \dfrac{?}{?}$

3) $\dfrac{\text{Cubic inch}}{\text{Cubic yard}} = \dfrac{?}{?}$

4) $\dfrac{\text{Kilometer}}{\text{Mile}} = \dfrac{?}{?}$

5) $\dfrac{\text{Number of degrees of all interior angles of an octagon}}{\text{Number of degrees of all interior angles of a pentagon}} = \dfrac{?}{?}$

6) $\dfrac{\text{Square foot}}{\text{Square mile}} = \dfrac{?}{?}$

7) $\dfrac{\text{Square foot}}{\text{Surface area of a cubic yard}} = \dfrac{?}{?}$

8) (Grams in a kilogram/legs on a spider) ÷ (Pounds in a ton/ounces in a pound) $= ?$

9) (Number of senators in the U.S. Senate) ÷ (Number of members in the U.S. House of Representatives) x 87 = ?

10) Atomic number of carbon ÷ Atomic number of argon ÷ (⅑) $= ?$

Chapter Nineteen
Analogies

Analogies are a fun way to learn. Look at the 3 analogies I put on the board. The first one I did for you.

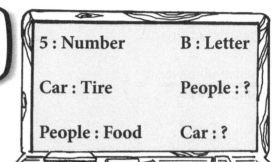

5 : Number   B : Letter

Car : Tire   People : ?

People : Food   Car : ?

Analogies are groups of words or numbers that are related in some way. The secret is to figure out how they are related. In the first analogy, "5" is a number and B is a letter. The word number refers to what the "5" is. The word letter refers to what "B" is. To solve the next analogy, we need to find out how a tire is related to a car?

I know how a tire is related to a car. It is what the car rides on. Now all I have to do is find what a person "rides" on. That is pretty easy. People use their feet to walk, so feet is the answer.

I get the idea. To solve the next one, I just ask myself what people use food for. They use it for energy so the answer to the third analogy is gas because cars use gas for energy.

There many areas of math where we can use analogies. Look at this one:

50% : ½   25% : ¼   10% : ?

I see how each one is related. Each percent is turned into a fraction, so the answer is ¹⁄₁₀ because 10% turned into a fraction is ¹⁄₁₀. Does anyone know why I am shouting?

50% : ½   25% : ¼   10% : ¹⁄₁₀

Maybe it is because you are so excited to finally learn how to do analogies. Try these two analogies:

Circle : π x radius x radius		Rectangle : ?
100 : 10	64 : 8	25 : ?

The first one gives the formula for area of a circle, so the answer must be length x width because that is how you find the area of a rectangle.

The second analogy was hard until I realized that 10 x 10 = 100 and 8 x 8 = 64. All I have to do now is ask myself what number multiplied by itself will equal 25. The answer of course is 5.

Circle : π x radius x radius		Rectangle : Length x width
100 : 10	64 : 8	25 : 5

Problem Set 1

Warmup: 50% : ½ : .5   25% : ¼ : ?

Level 1:   25% : ¼ : .25   20% : ⅕ : ?

Level 2:   33 ⅓% : ⅓ : .333   5% : ? : ?

Level 3:   75% : ¾ : .75   250% : 2 ½ : ?

Genius Level: 1% : ¹⁄₁₀₀ : .01   (¹⁄₁₀₀)% : ? : ?

Problem Set 2

Warmup: 6 sides : Hexagon   3 sides : ?

Level 1:   5 sides : Pentagon   8 sides : ?

Level 2:   Rectangle : Circle   Perimeter : ?

Level 3:   Rectangle : Length x Width   Circle : ?

Genius Level: Circle : Sphere   Rectangle : ?

Problem Set 3

Warmup: Yard : Foot : Inch   36 : 12 : ?

Level 1:   Circle : Sphere   Square : ?

Level 2:   81 : 9   144 : 12   10,000 : ?

Level 3:   Cubic foot : Cubic yard   1 : ?

Genius Level: Rectangle : Parallelogram   Square : ?

Problem Set 4

Warmup: 10 squared : 100   10 to the 4th : ?

Level 1:   10 : 5   5 : 2 ½   2 ½ : ?

Level 2: $20 : 2000 pennies $1,000,000 : ?

Level 3:   Cubic inch : Cubic foot   1 : ?

Genius Level: 12 : 4   4 : 1 ⅓   1 ⅓ : ?

Problem Set 5

Warmup: 15% : $^{15}/_{100}$ 98% : ?

Level 1:   75% : 15%   ¾ : ?

Level 2:   Rectangle : 2 lengths + 2 widths   Circle : ?

Level 3:   Cubic decimeter : cubic meter   1 : ?

Genius Level: Cubic millimeter : cubic meter   1 : ?

Problem Set 6

Warmup: $^{1}/_{10}$: 10%   ¼ : 25%   ½ : 50%   ¾ : ?

Level 1:   Earth's diameter : Earth's circumference   8000 miles : ?

Level 2:   Triangle : 180 degrees   Rectangle : ?

Level 3:   Trillion dollars : Million dollars   Million dollars : ?

Genius Level: $100 : 1 cent $1,000,000 : ?

Analogies
Level 1

1) ⅛ : ¼   ¼ : ?

2) Feet : 12 inches   Yard : ?

3) Minute : hour   1 : ?

4) 3 : 9   4 : 16   5 : 25   12 : ?

5) Gallon : quart   Quart : ?

Analogies
Level 2

1) ⅟₁ : ¼   ¼ : ⅟₁₆   ⅛ : ?

2) 5 minutes : 1 hour   5 seconds : ?

3) ⅟₁₆ : ¼   ⅛ : ?

4) 5,280 feet : mile   1,000 meters : ?

5) 75% : ¾   300% : ?

Analogies
Level 3

1)   ¹⁄₁₆ : ½   ⅛ : ?

2)   ¹⁄₁₀₀ : ¹⁄₁₀   ¹⁄₁₀ : ?

3)   Gallon : quart   Pint : ?

4)   1 second : 1 hour   1 : ?

5)   ¹⁄₁₀ : ¹⁄₅₀   10% : ?

Analogies
Genius Level

1)   100% : 50% : ½%   1 : .5 : ?

2)   Second : year   1 : ?

3)   2 : 8   3 : 27   4 : 64   10 : ?

4)   1 trillion : 1 million   1 million : ?

5)   Millimeter : kilometer   1 : ?

Chapter Twenty
Speed

I have a real dilemma. I am going on a 770 mile trip and cannot decide which road to take. One road allows me to travel at 70 miles per hour , but is very boring. The other is very scenic with many ocean views and redwood trees, but the speed limit is 55 miles per hour.

If we could figure out how much longer the scenic route would be, maybe that would help with your decision. Look at the calculation I did.

770 miles ÷ 70 miles per hour = 11 hours

770 miles ÷ 55 miles per hour = 14 hours

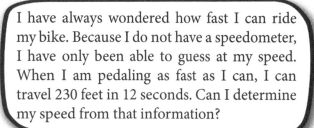

That sure made my decision easy. I can deal with boring to save 3 hours! While you are helping me with speed and math, I was hoping you could help me answer a question that has been bothering me for a long time.

I have always wondered how fast I can ride my bike. Because I do not have a speedometer, I have only been able to guess at my speed. When I am pedaling as fast as I can, I can travel 230 feet in 12 seconds. Can I determine my speed from that information?

That information is tremendous help! With a little thinking and logic, we can go from 230 feet per 12 seconds to the distance in an hour. Look at the 3 steps and pay close attention to the questions I ask myself.

Step 1:
230 feet per 12 seconds is how far in 60 seconds or one minute?
12 seconds x 5 = 60 seconds so we need to multiply 230 feet by 5 = 1150 feet per minute

Step 2:
1150 feet per minute so how far in one hour?
Multiply by 60
1150 feet x 60 minutes per hour =
69,000 feet per hour

Step 3:
69,000 feet per hour is equal to how many miles per hour?
Divide by 5280 because there are 5280 feet in one mile.   69,000 ÷ 5280 = 13.07 miles per hour

13 mph on a bike. Not too bad! Now that I know how to find miles per hour, I can find out my speed for the ¼ mile race I did in school. I ran the ¼ mile distance, which is .25 miles, in 50 seconds.

Step 1:
How many seconds in one hour?
60 seconds per minute x 60 minutes per hour = 3600 seconds in one hour.

Step 2:
How many 50 second parts are in 3600 seconds?
3600 seconds ÷ 50 seconds = 72 groups of 50 seconds in one hour

Step 3:
If I run ¼ mile every 50 seconds, how far will I run in 3600 seconds (one hour)?
There are 72 50-second groups in 3600 seconds so: .25 miles x 72 = 18 miles per hour

Here is another type of speed problem that is very fun. It has to do with the speed of trains.

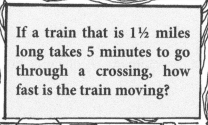

If a train that is 1½ miles long takes 5 minutes to go through a crossing, how fast is the train moving?

Even though the problem appears different, I believe it takes the same type of thinking and logic. I'll try this using steps like we did in the previous problems.

Step 1:
The train travels 1.5 miles in 5 minutes. How many groups of 5 minutes are in one hour?
60 minutes ÷ 5 minutes = 12 groups

Step 2:
If the train travels 1.5 miles per 5 minutes, how far will it travel if there are 12 groups of 5 minutes?
1.5 miles x 12 groups = 18 miles in one hour

1) If a car is traveling at 60 miles per hour, how long would it take to complete a trip that is 450 miles?

2) If a car travels ½ mile in 60 seconds, how many miles per hour is it traveling?

3) If a plane takes 15 seconds to travel one mile, what is the speed in miles per hour?

4) A speed of 2 miles per minute is equal to how many miles per hour?

5) A train traveling at 20 miles per hour takes 15 minutes to go through a crossing. How long is the train?

Problem Set 1

Warmup: If a car travels 100 miles in 2 hours, what is its speed in miles each hour?

Level 1: If a mosquito can fly one mile in 12 minutes, how far can it fly in one hour?

Level 2: A toddler walked ¹⁄₁₀ of a mile in 6 minutes. What was his speed in miles per hour?

Level 3: If a plane travels at a speed of one mile in 12 seconds, what is its speed in miles per hour?

Genius Level: Sound travels approximately 720 miles per hour. Light travels at a speed of 186,000 miles per second. What is light's speed in miles per hour?

Problem Set 2

Warmup: Sam the Snail's speed is 9 feet per hour. What is its speed in yards per hour?

Level 1: A speed of one mile each minute is equal to how many miles in an hour?

Level 2: Sam the Snail's brother's speed is 10 feet per hour. What is his speed in inches per minute?

Level 3: The speed of sound is approximately 1100 feet per second. How many seconds does it take sound to travel one mile? (Round to the nearest second.)

Genius Level: A chetah that runs at a speed of 60 miles per hour is how many times faster than a sprinter who runs 100 yards in 10 seconds?

a) Twice as fast   b) 3 times as fast   c) 5 times as fast   d) 10 times as fast

Problem Set 3

kilometer = 1000 meters
1 meter = 10 decimeters
1 decimeter = 10 centimeters
1 centimeter = 10 millimeters

Warmup: A pig can run at a speed of 18 kilometers per hour. How many meters can a pig run in an hour?

Level 1: An ostrich can run 65,000 meters in an hour. How many kilometers can an ostrich run in an hour?

Level 2: An ostrich can run 65,000 meters in an hour. How many millimeters can an ostrich run in an hour?

Level 3: A giant tortoise can travel 3 kilometers in 10 hours. How many meters will it travel in one hour?

Genius Level: A garden snail can move at a top speed of 500 decimeters per hour. What is its speed in kilometers per hour?

Problem Set 4

Warmup: A train that is 3 miles long took one hour to go by a railroad crossing. How fast was the train traveling?

Level 1: A train that is 4 miles long took 30 minutes to go by a railroad crossing. How fast was the train traveling?

Level 2: A train that is one mile long took 20 minutes to go by a railroad crossing. How fast was the train moving?

Level 3: A train that is 2 miles long took 6 minutes to go by a railroad crossing. How fast was the train moving?

Genius Level: A train is 1½ miles long and traveling at a speed of 45 miles per hour. How long would it take to travel through a railroad crossing?

Problem Set 5

Warmup: A snail crawls across the diameter of a 20 inch circle in 2 hours. What is its speed in inches per hour?

Level 1: A snail crawls around the 50 inch circumference of a circle in 4 hours. What is its speed in inches per hour?

Level 2: An ant crawls around a circle with a circumference of 60 inches in 12 minutes. What is its speed in inches per minute?

Level 3: A snail crawls around the circumference of a circle with a diameter of 50 inches in 2 hours and 37 minutes. How many inches per minute did the snail crawl?

Genius Level: A bug is riding on the point of the minute hand on a clock. If the minute hand is 6 inches long, what is the speed of the bug in feet per hour? (Round to the nearest whole number.)

Speed
Level 1

1) If a person jogs one mile in 12 minutes, what is her speed in miles per hour?

2) If it takes Emmanuel 20 minutes to walk the mile to his school, what is his speed in miles per hour?

3) A speed of 1000 millimeters per second is how many meters per second?

4) Snow started falling at 8:00 P.M. at a rate of 1½ inches per hour. At what time will the snow be ½ foot deep?

5) If it takes a car 4 hours to travel 240 miles, what is the average speed in miles per hour?

Speed
Level 2

1) A train that was 3 miles long took 20 minutes to pass through a railroad crossing. At what speed was it traveling?

2) If a horse can run at 20 miles per hour, how far will it travel in 3 minutes?

3) Car A is traveling at 70 miles per hour and is 20 miles behind Car B which is traveling at 60 miles per hour. How long until Car A catches up with Car B?

4) A tortoise and a hare are having a race. The tortoise runs at a speed of 1 mile per hour and the hare runs at 8 miles per hour. After ½ hour, how far ahead of the tortoise is the hare?

5) Snow starts falling at 7:30 A.M. at a rate of 1¼ inches every hour. At what time will the snow be 5 inches deep?

Speed
Level 3

1)   A train traveling at a speed of 20 miles per hour took 6 minutes to pass a large oak tree. How long was the train?

2)   An emu is running at a speed of 15 miles per hour. How far will it run in 8 minutes?

3)   A very fast meteor can travel at a speed of 18 miles per second. What is this speed in miles per hour?

4)   Many race cars can travel at a speed of one mile every 20 seconds. What speed is this in miles per hour?

5)   It takes sound 5 seconds to travel one mile through air. Sound can travel approximately 4.5 miles in 5 seconds through water. What is the speed of sound in miles per hour through water?

Speed
Genius Level

1) Apollo astronaut Eugene Cernan drove the Lunar Roving Vehicle at a record speed of 11.2 miles per hour. How many feet did he travel each second at this speed? (Round to the nearest whole number.)

2) A bug is riding on the tip of the hour hand of a clock. If the hour hand is 2 inches long, what is the speed of the bug in inches per hour? (Round to the nearest whole inch.)

3) A rabbit hops up a hill at 3 miles per hour and hops down the same hill at 6 miles per hour. What is the average speed of the rabbit on its trip up and down the hill? (Hint: It is not 4 ½ miles per hour.)

4) You spot a tyrannosaurus one mile away from you. If you can run at 10 miles per hour and the dinosaur can run at 30 miles per hour, how long until it catches you?

5) If a runner runs a ¼ mile race in 60 seconds, how many miles per hour did she run?

Chapter Twenty-One
Bases

I want to teach you about different bases, but before I do, I want to review the base system we use which is called base 10.

Look at the number 75,426. In base 10 this means:

6 ones or 6 x 1 = 6
+
2 tens or 2 x 10 = 20
+
4 hundreds or 4 x 100 = 400
+
5 thousands or 5 x 1000 = 5000
+
7 ten thousands or 7 x 10,000 = 70,000

Ten thousands	Thousands	Hundreds	Tens	Ones
7	5	4	2	6

70,00 + 5000 + 400 + 20 + 6 = 75,426

I remember how to find each column in base 10. All we do is multiply by 10 to find the next column.

So the next column would be 10 x 100,000 = 1,000,000. I think I understand base 10 and am ready for different bases.

| 100,000 | 10,000 | 1000 | 100 | 10 | 1 |

In base 10 you need to multiply by 10 to find each new column. What do you think you need to do for base 5?

It seems too easy, but my guess is that you multiply by 5.

| 15,625 | 3,125 | 625 | 125 | 25 | 5 | 1 |

If a visitor from the planet Quinton showed me the base 5 number 431, watch how I turn it into base 10.

15,625	3,125	625	125	25	5	1
				4	3	1

4 groups of 25 = 100
+
3 groups of 5 = 15
+
1 group of 1 = 1

100 + 15 + 1 = 116

That wasn't as scary as I thought. I can translate the base 5 number 222,222 using the same method.

15,625	3125	625	125	25	5	1
	2	2	2	2	2	2

2 groups of 3125 = 6250   **2 groups of 25 = 50**
+   **+**
2 groups of 625 = 1250   **2 groups of 5 = 10**
+   **+**
2 groups of 125 = 250   **2 groups of 1 = 2**

6250 + 1250 + 250 + 50 + 10 + 2 = 7,812

Find the first four columns for each base. The first one is done for you.

1) Base 4   **64   16   4   1**

2) Base 7

3) Base 9

4) Base 2

5) Base 8

Let's try one more problem translating from a different base into base 10. A person from the planet Tripod (who uses base 3) says he has a collection of 210 ants. How many ants would he have in base 10?

These are the columns in base 3. Now it is easy to turn the base 3 number 210 into base 10.

81	27	9	3	1
		2	**1**	**0**

2 groups of 9 = 18
            +
1 group of 3 = 3
            +
0 groups of 1 = 0

18 + 3 + 0 = 21

So now I know that he has 21 ants in our number system. If he said he had 20,021 ants, it would seem like he had a huge number of ants until I change 20,021 into base 10 and find that he only had 169 ants.

81	27	9	3	1
2	**0**	**0**	**2**	**1**

2 groups of 81 = 162
0 groups of 27 = 0
0 groups of 9 = 0
2 groups of 3 = 6
1 group of 1 = 1

162 + 0 + 0 + 6 + 1 = 169

Try these problems:

1) The base 3 number 11 is equal to _____ in base 10.

2) A visitor from the planet Octon says that he weighs 311 pounds. How much does he weigh in base 10?

3) Gabe is bragging that his allowance is $1000 per week in base 2. What is Gabe's allowance in base 10?

4) You can never have a base 7 number of 77. Why?

Changing From Base 10 to Other Bases

The next area we need to cover is translating from base 10 into different bases. Watch how I translate the base 10 number 158 into base 5 by using 5 simple steps.

158

Base 5 columns				
		(33)	(8)	(3)
625	125	25	5	1
	1	1	1	3

1) Are there any groups of 625 in 158? Of course there are not any, so we leave that column blank.

2) Are there any groups of 125 in 158? There is only one group with 158 - 125 = 33 left over. We will put the left over amount on top of the next column.

3) How many groups of 25 are in 33? There is one group of 25 with 33 - 25 = 8 left over. We will put the left over amount on top of the next column.

So the base 10 number 158 is 1113 in base 5. Using bases is almost like a secret code I can use with my friends.

4) How many groups of 5 are there in 8? One group with 8 - 5 = 3 left over. We will put the left over amount on top of the next column.

5) How many groups of one are in 3? There are 3.

Let's try one more to make sure you understand. I will translate $100 in base 10 into base 2. I think you will be surprised at how large a number $100 appears to be in base 2.

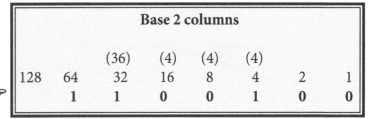

		(36)	(4)	(4)	(4)		
128	64	32	16	8	4	2	1
	1	**1**	**0**	**0**	**1**	**0**	**0**

Step 1: Are there any groups of 128 in 100? Of course there are not any, so we leave that column blank.

Step 2: Are there any groups of 64 in 100? There is only one group with 100 - 64 = 36 left over. We will put the left over amount on top of the next column.

Step 3: How many groups of 32 are in 36? There is one group of 32 with 36 - 32 = 4 left over. We will put the left over amount on top of the next column. There are no groups of 16 and no groups of 8 in 4 so we will put zeros there.

Step 4: How many groups of 4 are there in 4? One group with zero left over.

Step 5: Because there is nothing left, we will put zeros in the remaining columns.

Try the following problems:

1) $10 in base 10 is _____ in base 2.

2) 100 pounds in base 10 is _____pounds in base 9.

3) If you told a person from the planet Septon that you had 10 fingers, how many fingers would he think that you had?

4) Translate 10 fingers into base 7.

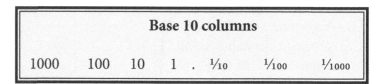

Base 10 columns

1000	100	10	1	.	¹⁄₁₀	¹⁄₁₀₀	¹⁄₁₀₀₀

Here is an advanced part of bases to try if you want a challenge. Decimals can be translated into different bases. Let's start by looking at base 10 decimal columns.

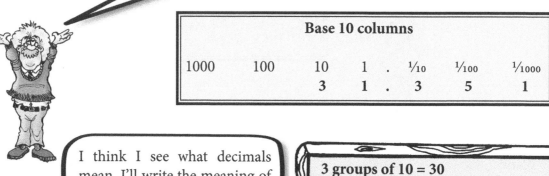

Base 10 columns

1000	100	10	1	.	¹⁄₁₀	¹⁄₁₀₀	¹⁄₁₀₀₀
		3	1	.	3	5	1

I think I see what decimals mean. I'll write the meaning of 31.351 on the board.

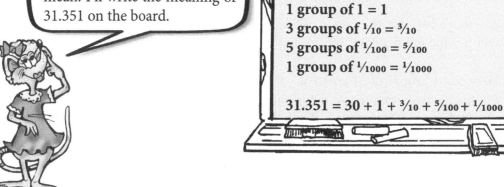

3 groups of 10 = 30
1 group of 1 = 1
3 groups of ¹⁄₁₀ = ³⁄₁₀
5 groups of ¹⁄₁₀₀ = ⁵⁄₁₀₀
1 group of ¹⁄₁₀₀₀ = ¹⁄₁₀₀₀

31.351 = 30 + 1 + ³⁄₁₀ + ⁵⁄₁₀₀ + ¹⁄₁₀₀₀

Write the first three decimal columns for each base. The first one is done for you.

1) Base 2   ½   ¼   ⅛

2) Base 5

3) Base 6

4) Base 9

5) Base 3

Problem Set 1

Warmup: In the number 7342, the 3 stands for 3 groups of 100 = 300. What does the 7 stand for?

Level 1: In base 6, the number 543 means something very different than 543 in base 10. In 543 base 6, the number 5 means 5 groups of 36 or 180. What does the number 4 stand for?

Level 2: The base 10 number 8 is turned into base 6 as shown below:

	Base 6 columns:	36	6	1
			1	2

How many groups of 36 are in 8? Obviously, there are no groups of 36 in 8 because 8 is too small.
How many groups of 6 are in 8? Answer: 1 group of 6 with 2 left over
How many groups of 1 are in the 2 left over? Answer: 2 groups of 1 are in 2.
Translate the base 10 number 23 into base 6.

Level 3: The first 6 columns for base 5 are: 3125   625   125   25   5   1

What are the first 12 columns for base 2?

Genius Level: It was mentioned to a person who uses base 2 that the United States has 100 senators. The person who uses base 2 drew a picture as shown below and asked: You have only this many senators?

x   x   x   x

Why did he think that the United States has a total of 4 senators?

Problem Set 2

Warmup: The base 10 columns are shown below. What is the missing column?

Base 10 columns: ?   10,000   1,000   100   10   1

Level 1: Fill in the missing base 5 column.

Base 5 columns: 625   125   ?   5   1

Level 2: Translate the base 5 number 121 into base 10.

Level 3: The base 10 number 37 is translated into base 5 as follows:

Base 5 columns: 25   5   1
                    1   2   2

How many groups of 25 are in 37?   Answer: 1 with 12 left over
How many groups of 5 are in 12?   Answer: 2 with 2 left over
How many groups of 1 are in 2? Answer: 2

Translate the base 10 number 99 into base 5.

Genius Level: A 5th grader from New York City visited Septon and thought the scale he stepped on was broken because he weighed 213 pounds. Translate the base 7 number 213 into base 10 so he will know how much he weighs.

Problem Set 3

Warmup: The base 5 number 100 means 1 group of _____.

Level 1: The base 5 number 42 means 4 groups of 5 plus 2 groups of 1 or 20 + 2 = 22
So we know that the base 5 number 42 is equal to 22 in base 10.
What is the base 5 number 24 equal to in base 10?

Level 2: The base 10 number 362 means (3 x 100) + (6 x 10) + (2 x 1). In base 5, what does the number 432 mean?

Level 3: Translate the base 5 number 341 into base 10.

Genius Level: A person using base 10 said that he drank .4 of his coffee. A person using base 5 said that she also drank .4 of her coffee. What fraction of each cup of coffee is left?

Fraction of base 10 person's coffee remaining: ?
Fraction of base 5 person's coffee remaining: ?

Problem Set 4

Warmup: The base 7 number 100 means 1 group of_____.

Level 1: The base 7 number 65 means 6 groups of 7 + 5 groups of 1. This means that 65 in base 7 is equal to 42 + 5 = 47 in base 10. What is the base 7 number 56 equal to in base 10?

Level 2: Translate the base 7 number 432 into base 10.

Level 3: The base 2 number 111 translated into base 10 is:
1 group of 4 + 1 group of 2 + 1 group of 1   4 + 2 + 1 = 7

Translate the base 2 number 10101 into base 10.

Genius Level: In base 10 the number 432.75 means: $400 + 30 + 2 + \frac{7}{10} + \frac{5}{100}$

What does the base 9 number 432.75 mean?

Problem Set 5

Warmup: The base 4 number 3,333 means 3 groups of 64 plus 3 groups of 16 plus 3 groups of_____ plus 3 groups of 1.

Level 1:   303 in base 10 means $(3 \times 100) + (0 \times 10) + (3 \times 1) = 303$

202 in base 4 means $(2 \times ?) + (0 \times ?) + (2 \times ?)$

Level 2: Translate the base 10 number 12 into base 4.

Level 3: I have 1,000,000 dollars in base 2. How much is that equal to in base 10?

Genius Level: A visitor from an island that uses base 2 said that his salary on the island is $1,000,000,000,000,000 per year. What is his salary in base 10?

Bases
Level 1

1) The number 32 means 30 + 2 = 32.   The number 482 means 400 + 80 + 2 = 482.   What does the number 8678 mean?

2) You are 9 years old and are going to write a number to show a person who uses base 7 how old you are. What number would you write?

3) A weight of 111 pounds in base 9 is _____ pounds in base 10.

4) An elementary student said that she is 1000 years old in base 2. How old is she in base 10?

5) If your father is 33 years old, how old is he in base 2?

Bases
Level 2

1) The base 10 number 35.1 means 30 + 5 + ¹⁄₁₀. If you were told that 35.1 is a base 7 number, it would mean:
        3 groups of ____
        5 groups of ____
        1 group of ____

2) A visiter from a base 5 island said that he weighed 400 pounds. Translate 400 base 5 to base 10.

3) A weight of 100 pounds in base 10 is____ in base 9.

4) Tell a base 5 person how many inches are in a foot.

5) Tell a base 9 person how many days it takes the Earth to travel around the sun (365 days).

Bases
Level 3

1) In the number 236.45, the 236 means 200 + 30 + 6. What does the .45 stand for?

2) You want to tell a person who uses base 5 how tall you are. If you are 4 feet and 3 inches, what would this be in base 5?

3) You want to tell a person who uses base 5 how tall you are. If you are 6 feet and 6 inches, what would this be in base 5?

4) Tell a base 2 person how many hours and minutes it takes the Earth to rotate once. (23 hours and 56 minutes)

5) Who has more money?

   Base 5 with $100   or   Base 2 with $10,000

Bases
Genius Level

1) The number 78.1 means $70 + 8 + \frac{1}{10}$. What does 365.0001 mean?

2) Bill has $1000 in base 10. Brianna has $1,000,000 in base 2. How much more money does Bill have than Brianna?

3) A giant cockroach weighs 28.5 ounces. What is this weight in base 2?

4) A person who uses base 2 asks you how much money you have in your pocket. If you have $5.75, what do you tell the base 2 person?

5) A person who uses base 9 mentions that the combination to her lock is 8503. Translate the combination to base 10.

Answers: Astronomy, Light and Sound

(Page 6)

1) ¾ of a pound

Half of 10 pounds is 5 pounds. Half of 1½ = ¾

2) 24 pounds

20 pounds is ⅕ of 100. ⅕ of 120 = 24

3) 19 kilograms

250 is 2.5 x 100 so weight in pounds is 17 x 2.5 = 42.5
42.5 x .45 = 19.125 kilograms

(Page 8)

1) 2 years old

Each 29 years is a Saturn year.

2) 13 years old

8 years x 365 days = 2920 days
2920 days ÷ 225 days in a Venus year = 12.98 or 13 years

3) One month old

7 years ÷ 84 years = $\frac{1}{12}$ of a year $\frac{1}{12}$ of a year = one month

(Page 9)

1) ½ mile

It takes sound 5 seconds to travel one mile, so it will travel ½ mile in 2 ½ seconds.

2) 11,160,000 miles

60 seconds x 186,000 miles per second = 11,160,000 miles

3) 12 miles

Sound travels one mile in 5 seconds. There are 12 five-second parts in 60 seconds.
60 ÷ 5 = 12 miles

Problem Set 1 (Page 10)

Warmup: 8.5 pounds

Because 50 is ½ of 100, divide 17 in half. $17 \div 2 = 8.5$

Level 1: 59 pounds

25 pounds is ¼ of 100 pounds $236 \div 4 = 59$ pounds

Level 2: 18 pounds

100 pounds = 15 pounds on the asteroid

20 pounds on Earth is ⅕ of 100 pounds so it would be ⅕ of 15 $15 \div 5 = 3$ pounds

Earth: 100 pounds + 20 pounds   Asteroid: 15 pounds + 3 pounds = 18 pounds

Level 3: 1890 kilograms

Because 100 pounds = 2800 pounds on the sun, a 150 pound monkey would weigh 1.5 x 2800 = 4200 pounds on the sun.   4200 pounds x .45 = 1890 kilograms

Genius Level: 263 pounds

Because there are 38 pounds on Mercury for each 100 pounds on Earth, each pound on Mercury would equal 100 ÷ 38 = 2.63 pounds on Earth.

100 pounds on Mercury x 2.63 = 263 pounds on Earth

Problem Set 2 (Page 11)

Warmup: 3,720,000 miles

186,000 miles per second x 20 seconds = 3,720,000 miles

Level 1: 37,200 miles

Speed of light: 186,000 miles each second ÷ 5 = 37,200 miles

Level 2: a) 1.3 seconds

Light travels 186,000 miles in one second

250,000 miles ÷ 186,000 miles per second = 1.3 seconds

Level 3: 31,536,000 seconds

There are 3600 seconds in an hour, therefore there are 24 x 3600 = 86,400 seconds in a day

86,400 seconds in a day x 365 days in a year = 31,536,000 seconds in a year

Genius Level: 5,865,696,000,000 miles

Light travels 186,000 miles each second.

Seconds in a year: 60 seconds in a minute x 60 minutes in an hour x 24 hours in a day x 365 days in a year = 31,536,000 seconds in a year.   31,536,000 seconds per year x 186,000 miles per second = 5,865,696,000,000 miles per year

Problem Set 3 (Page 12)

Warmup: ¼

> 2000 is ¼ of 8000 because 8000 ÷ 2000 = 4

Level 1: 11 times larger

> 89,000 ÷ 8000 = 11.125

Level 2: 41 minutes and 18 seconds

Mars day:	24 hours	37 minutes	22 seconds
Earth day:	23 hours	56 minutes	4 seconds

> Borrow an hour from the Mars day and make the minutes 97

Mars day:	23 hours	97 minutes	22 seconds
Earth day:	23 hours	56 minutes	4 seconds
	0 hours	41 minutes	18 seconds

Level 3: 1041.67 miles per hour

> 25,000 miles in 24 hours is 25,000 ÷ 24 = 1041.67 miles per hour

Genius Level: 8 miles per second

> Find the length of Jupiter's day in seconds:
> 9 hours x 3600 seconds per hour = 32,400 seconds
> 50 minutes x 60 seconds per minute = 3000 seconds + 28 seconds
> 32,400 + 3000 + 28 = 35,428 seconds
> 280,000 miles ÷ 35,428 seconds = 8 miles per second

Problem Set 4 (Page 13)

Warmup: 1 mile

> Lightning and thunder happen at the same time, but lightning goes to your eyes almost instantaneously while the sound of thunder travels 1 mile in the 5 seconds.

Level 1: ½ mile

> Sound travels 1 mile in 5 seconds, so it must travel ½ mile in 2 ½ seconds.

Level 2: 1 mile

> It takes sound 5 seconds to travel one mile, so sound will travel 2 miles in 10 seconds. The sound traveled 2 miles, but it had to go to the rock wall and back --- one mile each way.

Level 3: 1.6 miles

> Sound travels 1 mile in 5 seconds, so it travels 8 ÷ 5 = 1.6 miles in 8 seconds.

Genius Level: 18 miles per hour

> When thunder is heard 15 seconds after lightning, the storm is 3 miles away. When lightning and thunder occur at the same time, the storm is 0 miles away. The storm traveled the 3 miles in 10 minutes, which means it would travel 3 miles x 6 = 18 miles in 60 minutes.(Multiply by 6 because there are six 10 minute parts in an hour.)

Problem Set 5 (Page 14)

Warmup: 93,000 miles
> 186,000 miles/second ÷ 2 = 93,000

Level 1: 11,160,000 miles
> 186,000 miles per second x 60 seconds = 11,160,000 miles

Level 2: c) 7.5 times
> Earth's equator is 25,000 miles   Light travels 186,000 miles in a second.
> 186,000 ÷ 25,000 = 7.44 times

Level 3: 8 minutes and 20 seconds
> 93,000,000 ÷ 186,000 miles per second = 500 seconds for light to travel from the sun to
> the Earth.   500 seconds ÷ 60 seconds in a minute = 8 minutes and 20 seconds

Genius Level: d) 1,000,000 times faster
> It takes sound 5 seconds to travel one mile.
> Light travels 5 x 186,000 = 930,000 miles in 5 seconds
> Light travels 930,000 times faster than sound, so answer d) 1,000,000 is the best answer

Problem Set 6 (Page 15)

Warmup: 3 years old
> Each year on Jupiter is 12 years.   36 ÷ 12 = 3 years

Level 1: c) ¼ year old
> 165 years is a Neptune year so 41 is about ¼ of a Neptune year.

Level 2: 29 years old
> 18 years is 365 x 18 = 6570 days   6570 ÷ 225 days in a Venus year = 29.2 years

Level 3: 18.6 years old
> 35 years x 365 days = 12,775 days   12,775 ÷ 687 days in Mars year = 18.6 years

Genius Level: 10,512 days
> Jupiter year = 12 years x 365 = 4380 days x 24 hours per day = 105,120 hours
> 105,120 ÷ 10 hours per day = 10,512 Jupiter days.

Level 1 (Page 16)

1) 12 miles

Sound travels one mile in 5 seconds, so it travels 60 ÷ 5 = 12 miles in 60 seconds.

2) d) 200 pounds

A person who weighs 17 pounds on the moon weighs 100 pounds on Earth.
34 is twice 17   2 x 100 = 200

3) 390

585,000,000 ÷ 1,500,000 = 390

4) 3 miles

The lightning and thunder occur at the same time. Sound travels one mile in 5 seconds, so it would travel 3 miles in the 15 seconds it took for the sound to reach Belinda's ear.

5) 5.2 astronomical units

484,000,000 miles ÷ 93,000,000 = 5.2

Level 2 (Page 17)

1) 3 ½ miles

Sound travels one mile in 5 seconds, so it travels 3 ½ miles in 17 ½ seconds..

2) d) 4 times faster

Sound in air: 1100 feet per second
Sound in water: 4700 feet per second
4700 feet per second is about 4 times faster than 1100 feet per second.

3) c) ¹/₇ second

The Earth's 25,000 mile equator is a small fraction of the 186,000 miles sound travels in one second. 25,000 ÷ 186,000 is very close to ¹/₇.

4) 66,781 mph

585,000,000 ÷ 8760 = 66,780.8

5) d) 500 miles

7200 feet per day x 365 days in a year = 2,628,000 feet per year
2,628,000 feet per year ÷ 5280 feet per mile = 498 miles

Level 3 (Page 18)

1) b) 5 ½ hours

Pluto is approximately 3,675,000,000 miles from the sun.
3,675,000,000 miles ÷ 186,000 miles per second = 19,758 seconds
19,758 seconds ÷ 3600 seconds per hour = 5.48 hours

2) d) 1 trillion pounds

3) a) ¼₀₀

Earth to moon: 240,000 miles   Earth to sun 93,000,000
$^{240,000}/_{93,000,000} = ^{1}/_{387.5}$ which is closest to ¼₀₀.

4) 25,200 mph

3600 seconds in an hour x 7 miles per second = 25,200 miles per hour.

5) 6 mph

The storm started at 3 miles away because sound takes 5 seconds to travel each mile. 10 minutes later, the storm was 2 miles away because sound took 10 seconds to reach your ears. The storm traveled one mile in 10 minutes, which means it would travel 6 miles in 60 minutes.

Genius Level (Page 19)

1) c) 35 times as fast

Sound takes 5 seconds to go one mile so sound travels at a speed of ⅕ mile per second. The escape velocity is 7 miles per second.   ⅕ goes into 7 a total of 35 times

2) 30 mph

The storm was originally 6 miles away because 30 ÷ 5 seconds to travel one mile = 6 miles. 12 minutes later, the storm is directly overhead so the storm traveled 6 miles in 12 minutes. If it traveled 6 miles in 12 minutes, it would travel 30 miles in an hour. (There are five 12-minute parts in an hour.)

3) Sound: c) Almost to the top of the Empire State Building

The Empire State Building is 1454 feet tall and sound travels approximately 1100 feet per second.

Light: b) Almost to the moon

Light travels 186,000 miles in one second. The moon is approximately 240,000 miles from Earth.

4) c) 760 mph

Sound travels 1100 feet per second. There are 3600 seconds in an hour, so sound travels 3600 x 1100 = 3,960,000 feet in an hour.
3,960,000 feet ÷ 5280 feet per mile = 750 miles per hour.

5) 27.3 days

1,500,000 total miles ÷ 2288 miles per hour = 655.6 hours
655.6 hours ÷ 24 hours in a day = 27.3 days

Answers: Problem Solving

(Page 21)

1) 32 gallons

Each ¼ must equal 24 ÷ 3 = 8   Whole tank: 4 x 8 = 32

2) Saturday

If 5 days before yesterday is Saturday, we can label the boxes

5	4	3	2	1	Yesterday	Today	
Sat.	Sun.	Mon.	Tue.	Wed.	Thur.	Fri.	Sat.

3) $240

After drawing a picture, it is clear that the $80 remaining is ²⁄₆ of
the money. If $80 is ²⁄₆, then ¹⁄₆ = $40.   6 x $40 = $240

²⁄₆ = $80

(Page 23)

1) 8 miles

2 - 10 method:

If 2 inches on a map are equal to 10 miles, what is one inch equal to?

Now it is easy to see that one inch would equal 5 miles. You divided.

Real problem: 18 miles ÷ 2.25 = 8 miles

2) 1600 pieces

2 - 10 method: How many 2 pound pieces can be cut from a 10 pound chocolate bunny?

This is an easy problem: 10 ÷ 2 = 5   We need to divide.

Real problem: 1200 ÷ ¾   which is the same as: 1200 ÷ .75 = 1600

3) 2 x *n* + 4 x *t* or 2*n* + 4*t*

2 - 10 method: There are 2 ostriches and 10 horses on a farm. How many legs are there?

2 ostriches x 2 legs + 10 horses x 4 legs

Now we know we need to multiply *n* ostriches x 2 and *t* horses x 4

(Page 24)

1) Two hours

Think 1: If 2 people take 3 hours, then only one person would take 6 hours.

1 person: 6 hours;   2 people: 6 ÷ 2 = 3 hours;   3 people: 6 ÷ 3 = 2 hours

2) One hour

Think 1: Hose A in 1 hour = ½ tank; Hose B in 1 hour = ¼ tank;

Hose C in 1 hour = ¼ tank   ½ + ¼ + ¼ = 1

3) 3 bags

Think 1 cat: One bag feeds 2 cats for 7 days, so one bag will feed one cat for 14 days.

If one bag feeds one cat for 14 days, then 3 cats need 3 bags for 14 days.

Problem Set 1 (Page 25)

Warmup: Friday

Level 1: 16 gallons

Level 2: Wednesday

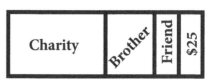

Level 3: $200

From the picture, it is easy to see that the $25 is ⅛ of the total. 8 x $25 = $200.

Genius Level: 16 gallons

Going from ⅜ full (which is ⁶⁄₁₆) to ⁹⁄₁₆ full is an increase of ⁹⁄₁₆ - ⁶⁄₁₆ = ³⁄₁₆.
³⁄₁₆ are equal to 3 gallons, so each ¹⁄₁₆ must be one gallon.
There are ¹⁶⁄₁₆ in the entire tank, so the tank holds 16 gallons.

Problem Set 2 (Page 26)

Warmup: $26

If chicken costs $2 per pound, how much does 10 pounds of chicken cost?
$2 x 10 pounds = $20
Real problem: $3.25 x 8 pounds = $26

Level 1: Twelve ¼ pound pieces

2-10 method: How many **2** pound pieces can be cut from a giant **10** pound chocolate
Easter Bunny? It is easy to see that the answer is 10 ÷ 2 = 5.
Real problem: 3 ÷ ¼ or How many ¼ pound pieces fit into a 3 pound piece?
There are four ¼ pound pieces in each whole, so there are 4 x 3 = twelve ¼ pound pieces in
the 3 pounds.

Level 2: 4n + 2p

2-10 method: If there are **2** pigs on a farm and **10** ducks, how many total legs are there?
Legs: 2 pigs x 4 legs each + 10 ducks x 2 legs each
Real problem: n pigs x 4 + p ducks x 2 n x 4 + p x 2   4n + 2p

Level 3: 696 rooms

2-10 method: If rooms **2** through **10** were cleaned, there would be 2, 3, 4, 5, 6, 7, 8, 9, 10
rooms cleaned. These add up to 9 rooms, so the rule must be subtract and add one.
(10 - 2 = 8 plus 1 = 9).
Real problem: 795 - 100 = 695 + 1 = 696

Genius Level: m/p

2-10 method: If it takes **2** minutes to read **10** pages, how long does it take to read one
page? If you are still unsure of what to do, switch the 2 and the 10:
If it takes **10** minutes to read **2** pages, how long does it take to read one page?
Answer: 10 ÷ 2 = 5 minutes.   Real problem: $m ÷ p$

Problem Set 3 (Page 27)

Warmup: $2.25

Level 1: ⅙

> 2-10 method: If there are **10** pizzas and **2** people, what fraction of a pizza would each person receive? Each person would receive 10 ÷ 2 = 5 pizzes.
> Real problem: 12 ÷ 72 = ⅙

Level 2: 16 miles

> 2-10 method: If **2** inches on a map are equal to **10** miles, how many miles is one inch equal to? 10 ÷ 2 = 5 miles.   Real problem: 12 miles ÷ .75 = 16 miles.

Level 3: 96 minutes

> 2-10 method: Juan can read **2** pages in one minute. How long will it take him to read **10** pages? 2 pages per minute means 10 ÷ 2 = 5 minutes to read 10 pages.
> Real problem: 72 ÷ ¾ = 96

Genius Level: *n* x *f* dollars

> 2-10 method: If the cost of each frog is **2** dollars, how much does it cost to buy **10** frogs?
> 2 x 10 frogs = $20.   Real problem: *n* x *f* dollars.

Problem Set 4 (Page 28)

Warmup: ¼

Level 1: ¹⁄₁₂

Level 2: 1 hour

> Think one hour:
> Daniel paints ½ of the fence in one hour
> Alicia paints ½ of the fence in one hour
> In one hour the whole fence is painted

Level 3: The whole fence

> Think one hour:
> Luke will paint ½ of the fence in one hour
> Laura will paint ¼ of the fence in one hour
> Jon will paint ¼ of the fence in one hour
> ½ + ¼ + ¼ = 1 whole fence in one hour

Genius Level: 1 hour and 20 minutes

> Think one hour:
> Hose A fills ¼ of the pool in one hour
> Hose B fills ½ of the pool in one hour
> In 60 minutes, both hoses fill ¾ of the pool, so each ¼ of the pool takes 20 minutes.
> There is ¼ of the pool left, so that quarter of the pool will take an extra 20 minutes to fill.

Problem Set 5 (Page 29)

Warmup: 30 hours   3 x 10 hours = 30 hours

Level 1: 2 hours

Think one person: 1 person: 2 x 3 = 6 hours   2 people: 3 hours   3 people: 2 hours

Level 2: 9 hours

Think one plow: 1 plow: 3 x 12 = 36 hours   2 plows: 36 ÷ 2 = 18 hours

3 plows: 36 ÷ 3 = 12 hours   4 plows: 36 ÷ 4 = 9 hours

Level 3: 6 hours

Think one person: 1 person: 3 x 8 = 24 hours   2 people: 24 ÷ 2 = 12 hours

If it takes 2 people 12 hours to paint 15 cars, then it would take them half as long to paint 7 ½ cars because 7 ½ is half of 15.

Genius Level: 5 hours

1 hose: 3 x (3 hours + 20 mintes) = 10 hours   2 hoses: 10 hours ÷ 2 = 5 hours

Level 1 (Page 30)

1) 6 hours

1 person: 3 x 8 = 24 hours   2 people: 24 ÷ 2 = 12 hour   4 people: 24 ÷ 4 = 6 hours

2) 15 gallons

⅕ of the tank must equal 3 gallons. 5 x 3 = 15 gallons

3) Tuesday

			Yesterday	Today	Tomorrow	
			Sunday	Monday	Tuesday	

4) 25

2-10 method: If **2** inches on a map are equal to **10** miles, how many miles is one inch equal to? 10 ÷ 2 = 5 miles

Real problem: 15 ÷ .6 = 25

5) ⅛ pizza

2-10 method: If **2** pizzas were bought for **10** people, how many would each person receive? 2 ÷ 10 = ⅕

Real problem: 12 ÷ 96 = ⅛

Level 2 (Page 31)

1) 14 hours

> 1 person: 7 x 10 = 70 hours
> 2 people: 70 ÷ 2 = 35 hours
> 5 people: 70 ÷ 5 = 14 hours

2) 18 gallons

> If ⅛ of the tank is 2 ¼ gallons, then ⅞ (which is the whole tank) are equal to 8 x 2 ¼ = 18

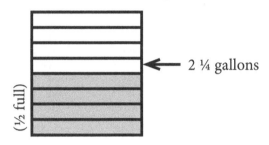

⟵ 2 ¼ gallons

(½ full)

3) Wednesday

		Yesterday	Today	Tomorrow	Day after Tomorrow	
		Wed.	Thursday	Friday	Saturday	

4) 4p + 2y legs

> 2-10 method: There are **2** pigs and **10** chickens in a truck. How many legs are in the truck?
> 2 pigs x 4 legs each = 8 legs   10 chickens x 2 legs each = 20 legs   Total: 28 legs
>
> Real problem:
> 4 x *p* pig legs = 4*p* legs   2 x *y* chicken legs = 2*y* legs   Total: 4*p* + 2*y* legs

5) 8n dollars

> 2-10 method: The cost of each baseball is **2** dollars. What is the cost of **10** baseballs?
> $2 x 10 = $20
>
> Real problem: $8 x *n* = 8*n* dollars

Level 3 (Page 32)

1) 2 hours and 30 minutes

6 hours and 40 minutes is equal to 400 minutes
3 people: 400 minutes
1 person: 3 x 400 = 1200 minutes
2 people: 1200 ÷ 2 = 600 minutes
8 people: 1200 ÷ 8 = 150 minutes or 2 hours and 30 minutes

2) 14 gallons

2 ⅝ gallons are equal to ³/₁₆ of the tank
2 ⅝ = 2⅛ gallons are equal to ³/₁₆ of the tank
Each ¹/₁₆ must then equal 2⅛ ÷ 3 = ⅞ gallons

16 x ⅞ = 14 gallons

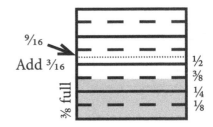

3) 90 minutes

2-10 method: If Jacob reads **2** pages per minute, how long will it take him to read **10** pages?
10 pages ÷ 2 per minute = 5 minutes

Real problem: 126 ÷ 1.4 = 90 minutes

4) 4.5 miles

2-10 method: The sound from a thunderstorm travels approximately **2** miles in one second. How far will sound travel in **10** seconds?
2 x 10 = 20 miles (Multiply to find the answer)

Real problem: ⅕ x 22.5 or .2 x 22.5 = 4.5 miles

5) 20 posts

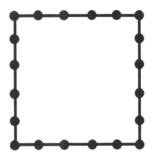

Genius Level (Page 33)

1) 2 hours

Abe does ⅙ of the driveway in one hour
Lana does ⅓ or $\frac{2}{6}$ in one hour
⅙ + $\frac{2}{6}$ or ½ the driveway in one hour, so two hours for the whole driveway

2) 6 days

1 person: 8 x 30 = 240 days
2 people: 240 ÷ 2 = 120 days
40 people: 240 ÷ 40 = 6 days

3) $2y/n$

2-10 method: If **2** airplanes can be built in **10** days, how long does it take to build 2 airplanes?
10 days ÷ 2 planes = 5 days per plane

Real problem: y days ÷ n airplanes = y/n per plane   2 planes: 2 x y/n or $2y/n$

4) 25 gallons

¾ full is 75% full
6 gallons make it 99% full,
so the 6 gallons must equal 24% of the tank.
(75% + 24% = 99%)

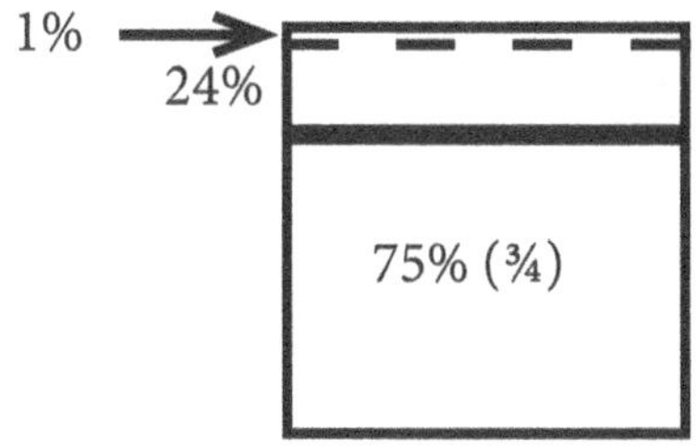

If 6 gallons are 24%, then each gallon is 24 ÷ 6 = 4%
There are 25-4% parts in 100% so the tank holds 25 gallons

5) m/p minutes

2-10 method: Lynn can type **2** pages in **10** minutes. How long does it take Lynn to type one page?

10 minutes ÷ 2 = 5 minutes per page

Real problem: m minutes ÷ p pages = m/p

Answers: Sequences, Place Value and Bases

(Page 35)

1) **15, 9, 3, -3, -9**
   The next number is found by subtracting 6. -3 -6 = -9

2) **¹⁄₁₆, ⅛, ¼, ½, 1**
   The next number is found by multiplying by 2. ½ x 2 = 1

3) **125, 25, 5, 1, ⅕**
   The next number is found by dividing by 5. 1 ÷ 5 = ⅕

(Page 35)

1) **10,000 column: 4   1000 column: 2**

2) **90,000 + 9000 + 900 + 90 + 9**

3) **8 is in the 100,000,000 column   9 is in the billion's column.**

(Page 37)

1) **4 groups of 343 = 1372 in base 10**

2) **(5 x 49) + (5 x 7) + (5 x 1) = 245 + 35 + 5**

3) **6 groups of 2401 = 6 x 2401 = 14,406 marbles in base 10**

(Page 39)

1) **1 group of 16**

2) **(1 x 16) + (1 x 4) + (1 x 1) = 21**

3) **$64**
   1,000,000 in base 2   The 1 is in the 64's column

Problem Set 1 (Page 40)

Warmup: -15
   Subtract 15 each time

Level 1: -80
   Subtract 95 each time

Level 2: 5 ⅝
   Divide by two each time

Level 3: ¹⁄₁₀₀
   Divide by 10 each time

Genius Level: 441
   $25 \times 25 = 625$, $24 \times 24 = 576$, $23 \times 23 = 529$, $22 \times 22 = 484$, $21 \times 21 = 441$

Problem Set 2 (Page 40)

Warmup: 9
   Add 1 ½ each time

Level 1: 6.25
   Each number is half of the number before it. Half of 12.5 = 6.25

Level 2: ¹⁄₃₂
   Each number is half the number before it and decimals and fraction are alternated.
   ½, ¼ , ⅛, ¹⁄₁₆, ¹⁄₃₂

Level 3:   10,000
   $1 \times 1 = 1$
   $2 \times 2 = 4$
   $3 \times 3 = 9$
   $4 \times 4 = 16$
   $100 \times 100 = 10,000$

Genius Level: 125
   $1 \times 1 \times 1 = 1$
   $2 \times 2 \times 2 = 8$
   $3 \times 3 \times 3 = 27$
   $4 \times 4 \times 4 = 64$
   $5 \times 5 \times 5 = 125$

Problem Set 3 (Page 41)

Warmup: -1

Subtract 4 each time   3 - 4 = -1

Level 1: 12 ½

Divide by 4 to get the next number

Level 2: One thousand

Divide by 1000 each time

Level 3: 1000, 1, $\frac{1}{1000}$

Divide by 1000 each time

Genius Level: 1601

1st number: $1 \times 1 + 1 = 2$ 2nd number: $2 \times 2 + 1 = 5$
3rd number: $3 \times 3 + 1 = 10$ 4th number: $4 \times 4 + 1 = 17$
40th number: $40 \times 40 + 1 = 1601$

Problem Set 4 (Page 41)

Warmup: 1000

Level 1: 6 groups of 10,000

Level 2: 343

Multiply by 7 to get the next column

Level 3: 1, 12, 144, 1728

First column is 1, then multiply by 12 to find each additional column

Genius Level: 1000

These numbers are the counting numbers in base 2.
Base 2 columns: 64, 32, 16, 8, 4, 2, 1
1: 1 group of 1 = 1
10: 1 group of 2 = 2
11: 1 group of 2 and 1 group of 1 = 3
100: 1 group of 4 = 4
101: 1 group of 4 and 1 group of 1 = 5
110: 1 group of 4 and 1 group of 2 = 6
111: 1 group of 4, 1 group of 2 and 1 group of 1 = 7
1000: 1 group of 8

Problem Set 5 (Page 42)

Warmup: 1, 6, 36, 216

Level 1: 4 ½

Add ⅞ each time

Level 2: ⁹⁄₁₀

Each number is found by dividing by 10

Level 3: .125   .0625   .03125

Divide by 2 to get the next number

Genius Level: 55

Each number is found by adding the previous 2 numbers

Level 1 (Page 43)

1) $1.25

Divide the money by 2 to get the next number $2.50 ÷ 2 = $1.25

2) 2 ¼

Add ½ to get the next number

3) 5 groups of 1000 or 5000

4) 2.25

Add .50 each time

5) 25

Each number is a perfect square

Level 2 (Page 43)

1) 8 groups of 100,000 or 800,000

2) 0

Subtract 11 each time

3) 32, 16, 8

Each column is found by multiplying by 2.

4) -7

Subtract 7 each time   0 - 7 = -7

5) 100

1st number: 1 x 1 = 1   2nd number: 2 x 2 = 4   3rd number: 3 x 3 = 9

Find each number by multiplying by itself: 10th number: 10 x 10 = 100

Level 3 (Page 44)

1) ¼

Divide by 4 each time.

2) 63

The space between each number is 2, then 4, then 8, then 16. The next space must be 32.
31 + 32 = 63

Another way: Multiply previous number by 2 and then add 1.

2 x 1 + 1 = 3   2 x 3 + 1 = 7   2 x 7 + 1 = 15
2 x 15 + 1 = 31   2 x 31 + 1 = 63

3) 49

10 x 10 = 100   9 x 9 = 81   8 x 8 = 64   7 x 7 = 49

4) December 20

The day in December is multiplied by itself to see how large the frog grew.
20 x 20 = 400

5) December 10th

6th: $64   7th: $128   8th: $256   9th: $512   10th: $1024

Genius Level (Page 44)

1) 5040

Second number in the sequence is 2 x previous number: 2 x 1 = 2
Third number in the sequence is 3 x previous number: 3 x 2 = 6
Fourth number in the sequence is 4 x previous number: 4 x 6 = 24
Fifth number in the sequence is 5 x previous number: 5 x 24 = 120
Sixth number in the sequence is 6 x previous number: 6 x 120 = 720
Seventh number in the sequence is 7 x previous number: 7 x 720 = 5040

2) 81   ⅑   1/81

3) ¼   1   4   16

Multiply by 4 to find the next number

4) $1

Divide by 1000 each time

5) 17

This is a list of prime numbers. 17 is the next prime number.

Answers: Decimals

(Page 46)

1) $30 + 8 + \frac{3}{10} + \frac{4}{100}$

2) $500 + 10 + 2 + \frac{1}{10} + \frac{1}{100} + \frac{1}{1000}$

3) $125 \quad 25 \quad 5 \quad 1 \quad . \quad \frac{1}{5} \quad \frac{1}{25}$

4) $(1 \times 16) + (2 \times 4) + (3 \times 1) + \frac{3}{4} + \frac{1}{16}$

5) $32 \quad 16 \quad 8 \quad 4 \quad 2 \quad 1 \quad . \quad \frac{1}{2} \quad \frac{1}{4} \quad \frac{1}{8} \quad \frac{1}{16}$

Problem Set 1 (Page 49)

Warmup: $1.25
$$\$2.50 \div 2 = \$1.25$$

Level 1: 8 pieces
$$10 \div \$1.25 = 8$$

Level 2: $12.60
$$2.4 \times \$5.25 = \$12.60$$

Level 3: $28.35
The cost per foot is $20.25 \div 15 = \$1.35$ $\$1.35 \times 21 = \$28.35$

Genius Level: $84
$\$990 \div 6.875 = \$144$ per foot $\$144$ per foot $\div 12 = \$12$ per inch $\$12 \times 7$ inches $= \$84$

Problem Set 2 (Page 49)

Warmup: 6 for $2
6 for \$.35 each : 6 x \$.35 = \$2.10

Level 1: $2.88
$$12 \times .24 = \$2.88$$

Level 2: $33.50
$\$40.20 \div 12$ pounds $= \$3.35$ per pound   10 pounds x \$3.35 per pound $= \$33.50$

Level 3: $\frac{2}{10,000}$ of a cent
4 is $\frac{4}{100}$ of a cent   5 is $\frac{5}{1000}$ of a cent   2 is $\frac{2}{10,000}$ of a cent

Genius Level: $15.30
2 hours & 15 minutes = 135 minutes. How many 7.5 minute segments are in 135 minutes? 135 minutes $\div$ 7.5 = 18. There are 18 7.5 minute parts in 135 minutes. Each of the 18 parts lose .2 gallons of milk: Total loss is 18 x .2 =3.6 gallons. 3.6 gallons x \$4.25 per gallon = \$15.30

Problem Set 3 (Page 50)

Warmup: 6 hours
There are two .5 mile parts in each mile. 2 parts x 3 miles = 6 parts = 6 hours

Level 1: 13 miles
.25 is equal to ¼ so there are 4 .25 hour parts in each full hour.
The dog will walk 4 miles each hour or 13 miles in 3.25 hours.

Level 2: 17 coins
2.125 ÷ .125 = 17

Level 3: 17.85 miles
4 hours and 12 minutes is 4 ⅕ hours or 4.2 hours. 4.2 hours x 4.25 miles per hour = 17.85 miles

Genius Level: 3.5 miles
David will walk 6.25 hours each day x 6 days = 37.5 hours for the whole trip.
131.25 total miles ÷ 37.5 hours = 3.5 miles each hour

Problem Set 4 (Page 50)

Warmup: 90 + 9 + ⁹⁄₁₀

Level 1: ⁵⁄₁₀₀₀
Place values are 10   1   .   ¹⁄₁₀   ¹⁄₁₀₀   ¹⁄₁₀₀₀

Level 2: (6 x 8) + (4 x 1) + ²⁄₈ + ⁵⁄₆₄ = 48 + 4 + ²⁄₈ + ⁵⁄₆₄
Place values for base 8 are: 8   1   .   ⅛   ¹⁄₆₄

Level 3: 64 ¼
Place values for base 2:   64   32   16   8   4   2   1   .   ½   ¼

Genius Level: Bill has $21.40 which is $9.60 more than Stanley
Stanley: Base 5 columns: 5   1   .   ⅕   ¹⁄₂₅
(2 x 5) + (1 x 1) + ⁴⁄₅ = 11 ⁴⁄₅ or $11.80
$21.40 - $11.80 = $9.60

Problem Set 5 (Page 51)

Warmup: 66.75 miles

    $3 \times 22.25 = 66.75$

Level 1: 555.75 miles

    28.5 miles for each gallon x 19.5 gallons = 555.75 miles

Level 2: 15.4 gallons

    $\$57.75 \div \$3.75 = 15.4$ gallons

Level 3: $239.40

    $1530 \div 25.5 = 60$ gallons needed   Price per gallon is $3.99 $\$3.99 \times 60 = \$239.40$

Genius Level: $63.75

    Car A: 3000 miles ÷ 50 mpg = 60 gallons   60 gallons x $4.25 = $255
    Car B: 3000 miles ÷ 40 mpg = 75 gallons   75 gallons x $4.25 = $318.75
    $318.75 - $255 = $63.75

Problem Set 6 (Page 52)

Warmup: 1.5 degrees

    $212° - 210.5° = 1.5$ degrees

Level 1: 16.6 degrees

    $48.6 - 32 = 16.6$

Level 2: 4.2 degrees

    Normal body temperature is 98.6°F $102.8 - 98.6 = 4.2$

Level 3: 71.44 degrees

    $368.77 - 297.33 = 71.44$ degrees

Genius Level: 500 degrees

    Absolute zero is -459.67°F   -459.67 to zero is going up 459.67 degrees
    Add $40.7 + 459.67 = 500.37$ degrees

Level 1 (Page 53)

1) 7.5 inches

 2.5 + 2.5 + 1.25 + 1.25 = 7.5 inches

2) $37.50

 10 x $3.75 = $37.50

3) 21 miles

 12 mph means that each ¼ hour the bike will travel 3 miles so .75 or ¾ would be 9 miles. One hour is 12 miles so 1.75 hours = 12 + 9 = 21 miles

4) 400 + 10 + 2 + $\frac{7}{10}$

5) $4.44

 12 inches in a foot x 37 cents = 444 cents or $4.44

Level 2 (Page 54)

1) 60.6 degree temperature drop

 Dropping from 32 to 0 is 32 degrees   Dropping from 0 to -28.6 is a drop of 28.6 degrees 32 + 28.6 = 60.6 degree temperature drop

2) $46.50

 $3.75 x 12.4 gallons = $46.50

3) $6.06

 $99.99 ÷ 16.5 = $6.06

4) 400 + 30 + 9 + $\frac{8}{10}$ + $\frac{9}{100}$

5) 52.2 degrees

 212 - 159.8 = 52.2

Level 3 (Page 55)

1) 7.3 degrees

Dropping 2.2: 103.2 - 2.2 = 101 degrees   Rising 4.9 degrees: 101 + 4.9 = 105.9 degrees
105.9 - 98.6 (normal body temperature) = 7.3 degrees

2) $6.93

Car A: 18.4 x $3.85 = $70.84   Car B: 16.6 x $3.85 = $63.91 $70.84 - $63.91 = $6.93

3) 6000 + 300 + 7 + ³⁄₁₀₀ + ⁸⁄₁₀₀₀

4) 12 cents

$9.60 ÷ 16 ounces in a pound = 60 cents per ounce
.2 is ²⁄₁₀ or ⅕ of an ounce   60 cents ÷ 5 = 12 cents

5) 90 cents

Price per gallon of gas: $49.50 ÷ 12.5 = $3.96   Price per quart of gas: $3.96 ÷ 4 = $.99
Milk: $1.89 per quart - Gas at $.99 per quart = $.90 or 90 cents

Genius Level (Page 56)

1) 4 hours and 35 minutes

There are 20 quarts in 5 gallons
20 quarts ÷ .4 quarts = 50 so there are 50 .4 quart parts in 5 gallons
50 parts x 5.5 minutes to leak one part = 275 minutes   4 hours and 35 minutes

2) 23.56 pounds

What did you multiply .16 by to get 9.92? 9.92 ÷ .16 = 62   62 x .38 = 23.56

3) 300 + 30 + 3 + ³⁄₁₀ + ³⁄₁₀₀ + ³⁄₁₀₀₀ + ³⁄₁₀,₀₀₀ + ³⁄₁,₀₀₀,₀₀₀

4) 7:06 P. M.

It will take Tippy 3.195 miles ÷ .45 miles per hour = 7.1 hours
.1 hours or ¹⁄₁₀ of 60 minutes is 6 minutes so it will take Tippy the turtle
7 hours and 6 minutes. He will arrive at 7:06 P.M.

5) 19,738 meters

Change 29,035 feet to meters: 29,035 x .304 = 8826.64 meters
10.911 kilometers x 1000 = 10,911 meters   8826.64 + 10,911 = 19,737.64 meters

Answers: Money

(Page 57)

1) $2.15

        2 cents lowers the cost to $7.85 $10 - $7.85 = $2.15

2) $10.25

        7 cents lowers the cost to $9.75 $20 - $9.75 = $10.25

3) $4.10

        Change from the quarter is 25 cents - 15 cents = 10 cents

        Change from the 20-dollar bill and 5-dollar bill is $25 - $21 = $4

(Page 58)

1) 80 euros

        Fraction 1   Fraction 2

$$\frac{1 \text{ euro}}{\$1.25} \quad \times \quad 80 \quad = \quad \frac{? \text{ euros}}{\$100}$$

        1 euro x 80 = 80 euros

2) 7 euros

        Fraction 1   Fraction 2

$$\frac{35 \text{ euros}}{\$50} \quad (\times \text{ or } \div) \text{ ?} = \frac{? \text{ euros}}{\$10}$$

        $50 ÷ 5 = $10   35 euros ÷ 5 = 7 euros

3) ⅔ of a euro

        Fraction 1   Fraction 2

$$\frac{1 \text{ euro}}{\$1.50} \quad \times \quad ? = \frac{? \text{ euros}}{\$1}$$

This is a little more difficult. What did you do to $1.50 to get $1? You took ⅔ of it so ⅔ of 1 euro = ⅔ euro.

Problem Set 1 (Page 59)

Warmup: $225
>2.5 x $90 = $225

Level 1: $45
>There are 10 decimeters in a meter. 5 decimeters is half of a meter. $90 ÷ 2 = $45

Level 2: $9
>There are 100 centimeters in a meter. 10 centimeters is ¹⁄₁₀ of a meter. $90 ÷ 10 = $9

Level 3: 9 cents
>There are 1000 millimeters in a meter. 1 millimeter is therefore ¹⁄₁₀₀₀ of a meter.
>$90 ÷ 1000 = $.09 or 9 cents

Genius Level: 1 cent
>There are 1 billion nanometers in a meter.
>1 nanometer is therefore ¹⁄₁,₀₀₀,₀₀₀,₀₀₀ of a meter.
>Cost per nanometer: $10,000 ÷ 1,000,000,000 = $.00001 $.00001 x 1000 nanometers = $.01
>or 1 cent.

Problem Set 2 (Page 60)

Warmup: $7
>$21 ÷ 3 = $7

Level 1: 25 cents
>$5 ÷ 4 = $1.25 per scoop $4.50 ÷ 3 = $1.50 per scoop

Level 2: $9
>There are 3 half pound pieces in 1 ½ pounds of salmon.
>$13.50 ÷ 3 = $4.50 per half pound or 2 x $4.50 = $9 per pound

Level 3: $35
>Language of algebra:
>Shinguards: n
>Soccer ball: n + $10
>Equation: $2n + 10 = 60$ $2n = 50$ $n = 25$ soccer ball = 25 + 10 = $35

Genius Level: 80n dollars + $15
>If the car was rented 5 days, the charge would be 5 x $80 + 15 cents per mile.
>The car was rented for n days and driven 100 miles:
>n x $80 + 15 cents x 100 n x 80 = 80n 15 cents x 100 = $15   80$n$ dollars + $15

Problem Set 3 (Page 61)

Warmup: $60

15 gallons x $4 = $60

Level 1: $70

10 gallons would cost $35 so 20 gallons would cost $70

Level 2: $160

1000 miles ÷ 25 miles per gallon = 40 gallons of gas are needed

40 gallons x $4 per gallon = $160

Level 3: 460 miles

$87 would buy $87 ÷ $4.35 = 20 gallons of gas

20 gallons x 23 miles per gallon = 460 miles

Genius Level: $3800

Cost of gas for the pickup truck: 15,000 miles ÷ 12 miles per gallon = 1250 gallons used in a year.   1250 gallons x $4 = $5000

Cost of gas for the Prius: 15,000 ÷ 50 = 300 gallons of gas used in a year

300 gallons x $4 = $1200 $5000 - $1200 = $3800

Problem Set 4 (Page 62)

Warmup: 10 cents

$1.20 ÷ 12 = 10 cents

Level 1: $3.33

$9.99 ÷ 3 = $3.33

Level 2: $1.65

The 2 cents would lower the cost to an even $18.35, which is easier to subtract from $20.
$20 - $18.35 = $1.65

Level 3: $989.01

At $10 each, the cost would be $990. That is one cent too high for each shirt or 99 cents too high. $990 - 99 cents = $989.01

Genius Level: December 8th

December 1: $100	December 2: $50	December 3: $25
December 4: $12.50	December 5: $6.25	December 6: $3.12
December 7: $1.56	December 8: 78 cents	

Problem Set 5 (Page 63)

Warmup: $120

$175 - $55 = $120

Level 1: $5,000,000

$2 x 2,500,000 = $5,000,000

Level 2: ½ euro

If one euro is equal to 2 dollars, then each dollar must be equal to half a euro.

Level 3: 100 times

.01 dollars is one cent

.01 cents is ¹⁄₁₀₀ of a cent

Genius Level: ⁴⁄₅ of a euro or .80 euro

One euro is equal to 5 quarters, so each quarter is equal to ⅕ of a euro.

4 quarters are in a dollar 4 x ⅕ = ⁴⁄₅ or .8 euros

Level 1 (Page 64)

1) $88

$4 x 22 = $88

2) $6 per pound

$24,000 ÷ 4000 = $6

3) 25 cents

If 5 dozen eggs cost $15, then each dozen cost $15 ÷ 5 = $3. $3 ÷ 12 = $.25 or 25¢

4) $1.20

4 quarts in one gallon. $4.80 ÷ 4 = $1.20

5) $3.50

$63.00 ÷ 18 = $3.50

Level 2 (Page 65)

1) 80 cents

The 3 cents drops the cost of the cup of coffee to $1.20. $2 - $1.20 = 80 cents

2) $45

A 500 mile trip would use 500 ÷ 50 = 10 gallons of gas.   10 gallons x $4.50 per gallon = $45

3) 60 cents

There are five ¼ pound parts in 1 ¼ pounds of bananas so each ¼ pound part cost 75 cents ÷ 5 = 15 cents

If ¼ pound cost 15 cents, then one pound cost 4 x 15 = 60 cents.

4) $29,600

There are 16 ounces in a pound.   16 x $1850 = $29,600

5) $188

1 + 2 + 5 + 10 + 20 + 50 + 100 = 188

Level 3 (Page 66)

1) $3.20

2 x 3 ½ = 7 feet, so 7 feet of rope cost 2 x 80 cents = $1.60

If 7 feet of rope cost $1.60, then 14 feet must cost 2 x $1.60 = $3.20

2) $1110.96

Round all items up one cent: $1 + $10 + $100 + $1000 = $1111

Each item is rounded up one cent, so the total is 4 cents too high.

$1111 - 4 cents = $1110.96

3) 100 minutes

.01 dollars is equal to one cent.

4) 10 cents

$4 ÷ 40 = $.10 or 10 cents

5) $59,200,000

There are 16 ounces in a pound and 2000 pounds in a ton.

This means there must be 16 x 2000 = 32,000 ounces in a ton

32,000 x $1850 = $59,200,000

Genius Level (Page 67)

1) $560

Gas used by the Prius: 3000 miles ÷ 50 miles per gallon = 60 gallons

Gas used by the pickup truck: 3000 miles ÷ 15 miles per gallon = 200 gallons

Cost of Prius gas: 60 gallons x $4 = $240

Cost of pickup truck gas: 200 gallons x $4 = $800 $800 - $240 = $560

2) $1.25

Because there are four 20-cent parts in 80 cents, each 20 cents of Canadian money is equal to 25 cents of United States money. One dollar of Canadian money has five 20-cent parts, so one Canadian dollar is equal to 5 x 25 cents = $1.25 of United States money.

3) 100,000 minutes

.001 cents is $\frac{1}{1000}$ of a cent, so the charge for 1000 minutes would be one cent.

One dollar is 100 cents   1000 minutes per cent

One dollar would be the charge for 100 x 1000 minutes = 100,000 minutes

4) Glen's Gold Shop

There are 28 grams in an ounce, so Keith's Gold By the Metric is selling gold at 28 x $99 = $2772 per ounce

5) $4950

Find how many .6 mile pieces are in 792 miles.   792 ÷ .6 = 1320

One gallon each .6 mile piece, so 1320 gallons of gas are needed.

Cost: 1320 x $3.75 = $4950

Answers: Fractions

(Page 72)

1) ²⁄₉ of a pizza

2) ¹⁄₃₂

3) ⅛ pizza

Problem Set 1 (Page 73)

Warmup: ½ of a pizza

Level 1: ¼ of a pizza

Level 2: ⅙ of a pizza

Level 3: ⅗ of a pizza

Each pizza has 5 pieces that are ⅕ of a pizza. If there are 3 pizzas, there are 15 pieces that are ⅕ of the pizza. 15 pieces shared with 5 people = 3 pieces each

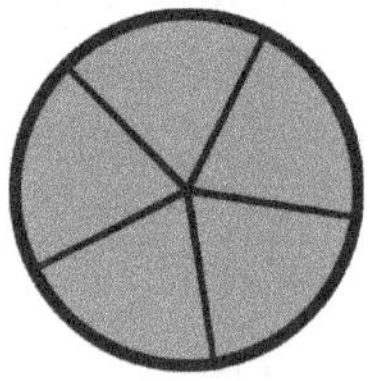

Genius Level: ⅛ of a pizza

Divide each whole pizza into 4 pieces.

There are 16 of the ¼ pizza pieces for the 4 pizzas.

There are two ¼ pieces in the ½ pizza.

Total ¼ pieces = 18   Because there are 36 people, each will have to share the ¼ pizza piece with another. Half of ¼ pizza is ⅛ of a pizza.

Problem Set 2 (Page 73)

Warmup: ¼

Divide ½ a pie into two equal pieces. The pieces are each ¼ of the pie.

Level 1: ⅛

Divide ½ of a pie into 4 pieces. They are each ⅛ of the pie.

Level 2: ¹⁄₁₆

Divide ¼ of a pie into 4 pieces. The pieces are each ¹⁄₁₆ of the pie.

Level 3: ¹⁄₁₆

½ of ½ is ¼   ½ of ¼ is ⅛   ½ of ⅛ = ¹⁄₁₆

Genius Level: ¹⁄₆₄

½ of ¼ = ⅛   ⅛ of ⅛ = ¹⁄₆₄

Problem Set 3 (Page 74)

Warmup: 1 ¼ pies

2 ½ pies divided by 2 = 1 ¼ pies

Level 1: 1/10

½ of 1/5 = 1/10

Level 2: 1/6

½ pie divided into three parts is equal to pieces that are each 1/6 of the pie.

Level 3: 1/12

⅓ divided into 4 parts would be 4 pieces that are each 1/12 of the cheese wheel.

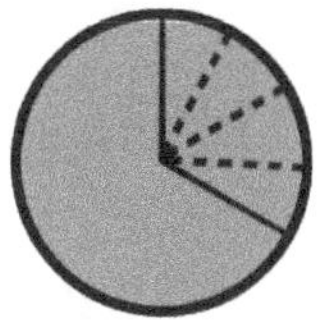

Genius Level: 2/15

⅔ of a cheese wheel has 10/15 of the entire cheese wheel.

10/15 split 5 ways = 2/15 each mouse

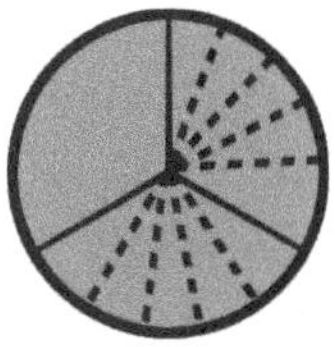

Problem Set 4 (Page 74)

Warmup: ⅛

Level 1: ⅛

¼ of the pie was left. ¼ divided into two equal pieces = two ⅛ pieces.

Level 2: 2 7/16

1 ⅜ is the same as 1 6/16 1 6/16 + 1 1/16 = 2 7/16

Level 3: 1/16

Dwight ate ½ or 8/16 of a pie while John ate ¼ or 4/16 of the pie. Lyndon ate ⅛ or 2/16 of the pie and Richard ate 1/16. How much of the pie is left? They ate 15/16 so 1/16 is left.

Genius Level: 1/48 of the candy bar

1/16 of the candy bar is left. 1/16 ÷ 3 = 1/48

Problem Set 5 (Page 75)

Warmup: ½ pound

Level 1: ⅝ pound

Level 2: ¹¹⁄₁₆ pounds

Level 3: 1 ¹⁵⁄₁₆ pounds

Genius Level: 1 ¹⁷⁄₂₀ pounds

Problem Set 6 (Page 76)

Warmup: 8 jumps

Level 1: 8 jumps   There are eight 1 ¼ pieces in 10 feet.

Level 2: 16 jumps   2 jumps on each short side and 6 on each long side

Level 3: 24 jumps   8 + 8 + 4 + 4 = 24 jumps

Genius Level: 16 jumps

   2 ¼ + 2 ⅛ + 2 + 1 ⅞ + 1 ¾ = 10   1 ⅝ + 1 ½ + 1 ⅜ + 1 ¼ + 1 ⅛ + 1 = 7 ⅞
   Total 11 jumps and a distance of 17 ⅞   5 more jumps: ⅞ + ¾ + ⅝ + ½ + ⅜ = 3 ⅛
   16 total jumps: 17 ⅞ + 3 ⅛ = 21 feet

Level 1 (Page 77)

1) ¼ pie

2) ¹⁄₁₀   Half of ⅕ is ¹⁄₁₀

3) ¹⁄₁₆   Half of ⅛ is ¹⁄₁₆

4) ⅛ pound   ¹⁄₁₆ pound would bring the right side to one pound. ¹⁄₁₆ more makes 1 ¹⁄₁₆

5) 8 jumps   Each 4 jumps = 3 feet   8 jumps = 6 feet

Level 2 (Page 78)

1) 8 pies   There are four ¼ pieces in each pie.

2) Twice as large   ⅜ is the same as ⁶⁄₁₆ which is twice as large as ³⁄₁₆.

3) ⅛   3600 ÷ 450 = 8   450 is ⅛ of 3600

4) 30 miles
      If ¼ inch = 12 miles, then ⅛ inch equals 6 miles
      If ⅛ inch = 6 miles, then ⅝ inch is 5 x 6 = 30 miles

5) ¹⁵⁄₁₆ pounds   5 - 4 ¹⁄₁₆ = ¹⁵⁄₁₆

Level 3 (Page 79)

1) $12

Karen has ½ + ¼ + ⅛ = ⅞ of the money she needs. She needs ⅛ more.   ⅛ of $96 = $12

2) ⅕

⅓ of ⅕ of a pie is ¹⁄₁₅ of the pie

3) ⅛ pizza

2 ½ pizzas have five ½ pieces

Each ½ piece has 4 pieces that are ⅛ each or 20 total pieces.

4) 5 ⅔ ounces

There are 16 ounces in a pound.   4 pounds x 16 = 64 ounces

¼ pound x 16 = 4 ounces   There are 68 ounces in 4 ¼ pounds.   68 - 62 ⅓ = 5 ⅔

5) $126

½ + ⅓ = ⅚ of the money was given away.   ⅙ of the money must be left

If $21 is ⅙ then Alicia started with 6 x $21 = 126

Genius Level (Page 80)

1) ¹⁄₁₂₀ of the day

How many 12 minute parts are in 24 hours?

60 minutes per hour ÷ 12 minutes = 5   There are five 12-minute parts per hour

24 hours per day x 5 = 120   So there are 120   12-minute parts in a day

2) 1650 pounds

Each ton = 2000 pounds   ⅛ ton = 2000 ÷ 8 = 250 pounds

There are 8250 pounds on the right side of the scale.

Each kilogram is equal to 2.2 pounds so: The left side of the scale weighs 3000 x 2.2 = 6600 pounds   8250 - 6600 = 1650 pounds

3) ¹⁄₁₀₀,₀₀₀,₀₀₀

$1,000,000 is equal to $1,000,000 x 100 = 100,000,000 pennies

1 penny/100,000,000 pennies

4) ¹⁄₄₈

½ of ⅓ = ⅙   ⅙ of ¼ = ¹⁄₂₄   ¹⁄₂₄ of ½ = ¹⁄₄₈

5) 0 yards. It is the end.

½ + ⅓ + ⅙ = 1 whole football field

Answers: Percents

(Page 85)

1) 2

$200 \div 100 = 2$

2) 250

25% means ¼   ¼ of 1000 = 250

3) 30

100% of 10 = 10   200% of 10 = 20   300% of 10 = 30

4) 1

10% means ¹⁄₁₀   ¹⁄₁₀ of 10 = 1

5) 16.15

$95 \div 100 = .95$ .95 x 17 = 16.15

6) 9

10% means ¹⁄₁₀ of 60 = 6   Additional 5% is half of 10% = 3

7) 150

75% means ¾   ¾ of 200 = 150

8) 8.4

10% of 80 = 8   1% is 80 ÷ 100 = .8   ½% is half of .8 = .4

9) 4

20% is ⅕   ⅕ of 20 = 4

10) 2.5

1% of 500 is 500 ÷ 100 = 5   ½% is half of 5 = 2.5

(Page 87)

1) $185.50

Sales tax: 6% = .06   .06 x $175 = $10.50 $175 + $10.50 = $185.50

2) $6.25

25% = .25   .25 x $25 = $6.25

3) $132

72.5% = .725   .725 x $480 = $348 $480 - $348 = $132

4) $17.29

19% = .19   .19 x $91 = $17.29

5) 25¢

Thinking way: 1% of $100 = $1 so ¼% must equal ¼ of $1 = 25 cents

(Page 88)

1) 45%

2) 8%

3) 11%

4) .5%

5) 325%

(Page 88)

1) 90%

45 (compared to) 50 = $^{45}/_{50}$ $^{45}/_{50}$ is 45 ÷ 50 = .90   .90 = 90%

2) 90%

3 wrong means 27 correct   27 (compared to) 30 = $^{27}/_{30}$ 27 ÷ 30 = .9   .9 = 90%

3) 1%

4 (compared to) 305 = $^{4}/_{305}$ 4 ÷ 305 = .013 .013 = 1.3%   Answer = 1%

4) 3%

.25 (compared to) 7.25 = $^{.25}/_{7.25}$ .25 ÷ 7.25 = .034   .034 = 3.4%   Answer = 3%

5) 21%

8 feet 11 inches = 8 x 12 = 96 + 11 = 107 inches
22 inches (compared to) 107 inches = 22 ÷ 107 = .2056   .2056 = 20.56%
Answer = 21%

(Page 89)

1) 50%

> Increase of 25 cents compared to original of 50 cents $^{25}/_{50} = .5 = 50\%$

2) 25%

> Increase of one foot compared to original of 4 feet $\frac{1}{4} = .25 = 25\%$

3) 75%

> Drop of $3 per week compared to original of $4 per week $\frac{3}{4} = .75 = 75\%$

4) 10%

> Drop of $2 per hour compared to original of $20 per hour $^{2}/_{20} = .10 = 10\%$

Problem Set 1 (Page 90)

Warmup: $15

> 1% of $100 is $1   5% of $100 is $5   10% of $100 is $10   15% of $100 is $15

Level 1: $2

> 10% of $10 is $^{1}/_{10}$ of $10 or $1.   20% is double that or $2
> Another method: Change 20% to a decimal and multiply   .2 x $10 = $2

Level 2: $7.50

> 10% of $50 is $^{1}/_{10}$ of $50 or $5 If 10% is $5, then 5% is half of that or $2.50
> $5 + $2.50 = $7.50
> Another method: Change 15% to a decimal and multiply   .15 x $50 = $7.50

Level 3: 30%

> If Elizabeth left a 10% tip, she would have left $9.
> $27 is 3 times $9 so she must have left a 30% tip.
> Another method: 27 compared to 90. $^{27}/_{90} = .30 = 30\%$

Genius Level: $12

> 100% of the meal price + 15% of the meal price = $92
> 115% of the meal price = $92 There are 23   5% parts of 115% (115% ÷ 5% = 23)
> Each 5% part is equal to $92 ÷ 23 = $4   15% tip is therefore 3 x $4 = $12

Problem Set 2 (Page 90)

Warmup: $8

> 1% of $100 is $1 so 8% is $8

Level 1: $2

> 10% of $50 is $50 ÷ 10 = $5   Each 1% is equal to $5 ÷ 10 = 50 cents
>
> 4% is = 4 x 50 cents = $2
>
> Another method: Change 4% to a decimal and multiply   .04 x $50 = $2

Level 2: $2.80

> 1% of $40 is $40 ÷ 100 = 40 cents   7 x 40 cents = 280 cents or $2.80
>
> Another method: .07 x $40 = $2.80

Level 3: $50.40

> Discount is 20%   10% of $60 is $6 so 20% must be $12   Sale price is $60 - $12 = $48
>
> Sales tax is 5% of 48   10% of $48 is $4.80 so 5% is half of that or $2.40
>
> $48 + $2.40 = $50.40   Another method: Discount is .20 x $60 = $12
>
> New price: $48   Sales tax is .05 x $48 = $2.40   Total cost: $48 + $2.40 = $50.40

Genius Level: $50

> 7% of something is $3.50 so 1% must be $3.50 ÷ 7 = 50 cents
>
> If 1% is 50 cents, then 100% is 100 x 50 cents = 5000 cents or $50
>
> Another method: 7% of something (n) = $3.50
>
> Change to decimal: .07 of n = $3.50 $n = 3.50 ÷ .07$ $3.50 ÷ .07 = 50$

Problem Set 3 (Page 91)

Warmup: 94%

Level 1: 40 questions

> 10% of 50 questions is 5 questions
>
> Because 80% is 8 times 10%, 80% of the questions must equal 8 x 5 questions or 40
> questions.   Another method: .80 x 50 = 40

Level 2: 60%

> Because there are 80 questions, each 8 correct answers would be 10%.
>
> How many 8's are in 48 correct questions?
>
> 48 ÷ 8 = 6 groups of 8 correct answers with each one being 10% 6 x 10% = 60%
>
> Another method: Percentage scores are found by comparing correct answers to the total
> number of questions.   48 correct ÷ 80 total questions = .60 or 60%

Level 3: 59 or more correct

> 1% of 64 questions is 64 ÷ 100 or .64   92% must be 92 x .64 = 58.88
>
> Another method: 92% = .92   .92 x 64 = 58.88 or 59 questions

Genius Level: 54 questions

> 87% of some number is 47   .87 x n = 47   Divide both sides by .87 $n = 54$

Problem Set 4 (Page 92)

Warmup: One package

1% of 100 is 1

Level 1: 30 packages

1% of 1000 is 1000 ÷ 100 = 10 packages   3% = 30 packages

Another method: Change 3% to a decimal   .03 x 1000 = 30

Level 2: 75 packages

10% of 500 = 50   5% is therefore half of 50 = 25   50 + 25 = 75

Another method: Change 15% to a decimal   .15 x 500 = 75

Level 3: ½%

If it lost 1% of its packages, it would lose 10,000 ÷ 100 = 100 packages.

50 is half of 100 so the company lost ½% of the packages.

Another method: 50 is what percent of 10,000   50 ÷ 10,000 = .005 or ½%

Genius Level: One package

Find 1% of 1,000,000   1,000,000 ÷ 100 = 10,000

.0001% is $\frac{1}{10,000}$ of a percent $\frac{1}{10,000}$ of 10,000 is equal to 1

Problem Set 5 (Page 92)

Warmup: $150

50% means ½   Half of $300 is $150

Level 1: $360

10% of $400 is $40 $400 - $40 = $360

Level 2: $30

A 20% discount is $\frac{1}{5}$ of 50% or $10. New price: $50 - $10 = $40

An additional 25% off $40: 25% means ¼

¼ of $40 is equal to another $10 discount. $40 - $10 = $30

Level 3: $99.50

½% off does not mean ½ off   1% of $100 is $1 so ½% of $100 is 50 cents

$100 - 50 cents = $99.50

Genius Level: $524.88

Monday: $800

Tuesday: $720   10% of $800 is $80 $800 - $80 = $720

Wednesday: $648   10% of $720 = $72 $720 - $72 = $648

Thursday: $583.20   10% of $648 = $64.80 $648 - $64.80 = $583.20

Friday: $524.88   10% of $583.20 = $58.32 $583.20 - $58.32 = $524.88

Problem Set 6 (Page 93)

Warmup: 1%

The change is $1   Compare $1 to $100 $1 is 1% of $100

Level 1: 50% increase

Percent of increase or decrease: (Amount of change) ÷ (The starting number)
Change is $5 $5 ÷ $10 = .50 or 50%

Level 2: 80% decrease

(Amount of change) ÷ (The starting number)
Started at $75 $60 change   ($75 - $15 =$60) $60 ÷ $75 = .80 or 80%

Level 3: 300%

Percent of increase or decrease: (Amount of change) ÷ (The starting number)
Started at $15,000   Increase was $45,000   ($60,000 - $15,000)
45,000 ÷ $15,000 = 3   3 as a percent is 300%

Genius Level: $11.25

8 years old: Decreased 50%   Cut in half: $40 ÷ 2 = $20
9 years old: Increased 50%   50% of $20 is $10 $20 + $10 = $30
10 years old: Decreased 50%   Cut in half: $30 ÷ 2 = $15
11 years old: Increased 50%   50% of $15 is $7.50 $15 + $7.50 = $22.50
12 years old: Decreased 50%   Cut in half: $22.50 ÷ 2 = $11.25

Problem Set 7 (Page 94)

Warmup: 5%

$5 is 5% of $100

Level 1: $8

1% of $400 is $4   If 1% is $4, then 2% is $8   Another method: .02 x $400 = $8

Level 2: $16

The difference in interest rates is 7% - 5% = 2%   1% of $800 is $8 so 2% of $800 is $16

Level 3: $1060.90

Second year: 3% of $1030   1% of $1030 = $10.30 so 3% must be 3 x $10.30 = $30.90
$1030 + $30.90 = $1060.90

Genius Level: $2,048,000

Money will double every 72 ÷ 12 = 6 years

6 years: $2000   12 years: $4000   18 years: $8000   24 years: $16,000
30 years: $32,000   36 years: $64,000   42 years: $128,000   48 years: $256,000
54 years: $512,000   60 years: $1,024,000   66 years: $2,048,000

Problem Set 8 (Page 95)

Warmup: 2 ½ inches

50% means half. Half of 5 inches is 2 ½ inches

Level 1: $200

25% means ¼   ¼ of $800 is equal to $200

Level 2: $500

Find 5% of $10,000   .05 x $10,000 = $500

Thinking method: 10% of $10,000 is $\frac{1}{10}$ of $10,000 or $1000

5% is half of 10% so   5% of $10,000 is half of the $1000 or $500

Level 3: $5

1% of $1000 is equal to $1000 ÷ 100 = $10

If 1% is equal to $10 then ½% is equal to $5

Genius Level: $286.25

22.9% of $15,000 is .229 x $15,000 = $3435   3435 ÷ 12 months = $286.25

Problem Set 9 (Page 95)

Warmup: 1 penny

One dollar is 100 cents   1% of 100 cents is one cent

Level 1: $32

Find 40% of $80   .40 x $80 = $32

Thinking method: 10% of $80 is $8   40% is 4 x $8 = $32

Level 2: $135

100% of Stu's allowance is $90   50% of Stu's allowance is $45

100% + 50% = 150% $90 + $45 = $135

Level 3: $5.25

17 ½% is 17.5% or .175   .175 x $30 = $5.25

Thinking method: 1% of $30 is $30 ÷ 100 = 30 cents

If 1% of $30 is 30 cents, then ½% is 15 cents

17 x 30 cents = $5.10   + ½% or 15 cents = $5.25

Genius Level: $738.80

The total amount of taxes that are deducted: 6.2% + 1.45% = 7.65%

Find 7.65% of $800   .0765 x $800 = $61.20 $800 - $61.20 = $738.80

Level 1 (Page 96)

1) $10

Thinking method: 5% of $100 is $5   so   5% of $200 must be 2 x $5 = $10

Alternative method: .05 x $200 = $10

2) 88%

If Jacob got 12 out of 100 wrong, then he correctly answered 100 - 12 = 88 questions

3) $40

10% of $50 is $50 ÷ 10 = $5   so   20% is 2 x $5 = $10

Discount is $10 $50 - $10 = $40

4) $1.50

Thinking method: 10% of $10 = $1   so   5% must be 50 cents

15% is therefore $1 + 50 cents = $1.50

Alternative method: .15 x $10 = $1.50

5) 50%

Percent of increase is the amount of increase divided by what the number started at.

$4 (increase) ÷ $8 (started at) = ½   ½ = 50%

Level 2 (Page 97)

1) $47.50

Thinking method: 10% of $50 is $5   5% must be half of $5 or $2.50

$50 - discount of $2.50 = $47.50

Alternative method: Discount is .05 x $50 = $2.50

2) $17.50

Thinking method: 7% of $100 = $7   so   7% of $50 = $3.50

The dog cost $250 so the sales tax is $7 + $7 + $3.50 = $17.50

Thinking method number 2: 1% of $250 is $250 ÷ 100 = $2.50   7% is 7 x $2.50 = $17.50

3) $22.50

Thinking method: 10% of $150 = $15   so 5% is $15 ÷ 2 = $7.50 $15 + $7.50 = $22.50

Alternative method: .15 x $150 = $22.50

4) $162

Thinking method: 1% of $900 = $9   so   18 x $9 = $162

Alternative method: .18 x $900 = $162

5) $60

Thinking method: 1% of $2000 = $20   ($2000 ÷ 100)   3% is 3 x $20 = $60

Alternative method: .03 x $2000 = $60

Level 3 (Page 98)

1) $49.50

½ of $100 is $50   1% of $100 = $1

½% is therefore $1 ÷ 2 = 50 cents $50 - 50 cents = $49.50

2) 15%

Thinking method: If the $200 coat was discounted 10%, it would be discounted

$200 ÷ 10 = $20   Each 10% is $20 so each 5% must be $10   10% + 5% = 15%

Alternative method: Percent of discount is change ÷ original price

change is $30   original price is $200   30 ÷ 200 = .15 or 15%

3) $12.50

Monday: 50% of $100 = $50 $100 - $50 = $50 left

Tuesday: 50% of 50 = $25 $50 - $25 = $25

Wednesday: 50% of $25 = $12.50 $25 - $12.50 = $12.50

4) $6.25

Each person would leave a 5% tip

10% of $125 is $125 ÷ 10 = $12.50   5% is $12.50 ÷ 2 = $6.25

5) $500

15% of Eric's pay is $75   so   5% of Eric's pay is $75 ÷ 3 = $25

If 5% is $25, then 10% must be twice that or $50.

If $50 is 10%, then the whole paycheck must be 10 x $50 = $500

Genius Level (Page 99)

1) ¼%

There are 530 + 410 + 257 + 3 = 1200 marbles. 1% of the marbles would be

1200 ÷ 100 = 12 marbles, so 3 marbles are ¼ of that or ¼% of the marbles.

2) 6 questions wrong

1% of 87 questions is 87 ÷ 100 = .87

If 1% is equal to .87 questions, then 93% is equal to 93 x .87 = 80.91 questions correct

Nick must get at least 81 correct, which is 6 incorrect.

3) $5.48

10% of $10,000 is $1000 so 20% must be $2000 per year. $2000 ÷ 365 = $5.48

4) $727.60

10% discount would be $80, so a 15% discount must be $120. Discounted price of the

guitar is $800 - $120 = $680. 1% sales tax would be $680 ÷ 100 = $6.80, so

7% = 7 x $6.80 = $47.60 $680 + $47.60 = $727.60

5) 700%

Percent of increase is the amount of increase ÷ original price

Increase in price is $14,000 ÷ by the $2000 original price = 7 or 700%

Answers: Translating Decimals, Fractions and Percents

(Page 102)

	Percent	Fraction	Decimal
1)	35%	$7/20$	.35
2)	25%	$1/4$	.25
3)	5%	$1/20$	.05
4)	1%	$1/100$	.01
5)	100%	1	1
6)	250%	$2\ 1/2$	2.5
7)	1000%	10	10
8)	12.5%	$1/8$	.125
9)	9%	$9/100$	.09
10)	8.5%	$85/1000 = 17/200$	.085
11)	.25%	$1/400$	.0025
12)	58.33%	$7/12$	.5833
13)	2%	$1/50$	.02
14)	70%	$7/10$	.7
15)	26.4%	$33/125$	.264

Problem Set 1 (Page 103)

Warmup: ¼

75% is equal to ¾ so ¼ of the marbles are left.

Level 1: 25%

¼ the bag must be blue   ¼ = .25 = 25%

Level 2: ⅕

If 80% are red, then 20% are green. 20% = $20/100$ = ⅕

Level 3: 5%

½ = 50%   ⅕ = 20%   ¼ = 25%

50% + 20% + 25% = 95% passed the test so 5% failed

Genius Level: 81.25%

⅛ + $1/16$ = $3/16$ 1 - $3/16$ = $13/16$ 13 ÷ 16 = .8125 or 81.25%

Problem Set 2 (Page 104)

Warmup: 25%

¼ is equal to 25%

Level 1: 20%

$\frac{1}{5} = .20 = 20\%$

Level 2: ⅓ and 33 ⅓%

12 inches is ⅓ of 36 inches   ⅓ = .333 = 33 ⅓%

Level 3: 500%

Building B's height is 5 times the height of Building A's height.

2 times something is 200%   3 times something is 300%   5 times something is 500%

Genius Level: ¹⁄₁₀₀₀ and ¹⁄₁₀%

There are 1000 millimeters in a meter, so a millimeter is ¹⁄₁₀₀₀ of a meter.

¹⁄₁₀₀₀ = .001 = .1% or ¹⁄₁₀%

Problem Set 3 (Page 105)

Warmup: 3 feet 1 inch

50% means half   Half of 6 feet 2 inches = 3 feet 1 inch

Level 1: $8 per week

100% of $8 is $8

Level 2: $32 per week

400% of $8 means 4 times $8 = $32

Level 3: 325%

3 ¼ = 3.25 = 325%

Genius Level: $12.80 per week

250% of something = $32   2.5 x something = $32 $32 ÷ 2.5 = $12.80

Problem Set 4 (Page 106)

Warmup: 60 cents

$1 ÷ 5 = 20 cents   20 cents x 3 = 60 cents

Level 1: $3/20$

15% = $^{15}/_{100}$ = $^{3}/_{20}$

Level 2: 67 cents

1 dollar = 100 cents   .50 dollars = 50 cents   .67 dollars = 67 cents

Level 3: One cent

.50 cents = ½ cent   2 apples cost ½ cent so 4 apples would cost 1 cent.

Genius Level: $^{1}/_{1250}$

.08% = .0008 = $^{8}/_{10,000}$ = $^{1}/_{1250}$

Problem Set 5 (Page 107)

Warmup: .500

Batting average is a fraction with hits on top and official plate appearances on the bottom.
$^{10}/_{20}$ = ½ = .500

Level 1: 25

A batting average of .250 means $^{250}/_{1000}$ or ¼ or one hit for every 4 times at bat.
100 times at bat = 25 hits

Level 2: 45

.200 = $^{1}/_{5}$ A hit $^{1}/_{5}$ of the time   225 ÷ 5 = 45

Level 3: .406

205 hits/505 at bats   205 ÷ 505 = .4059 or .406

Genius Level: .006

1 hit/165 at bats = .00606 = .006

Level 1 (Page 108)

	Percents	Fractions	Decimals
1)	25%	¼	.25
2)	75%	¾	.75
3)	10%	$^{1}/_{10}$	.1
4)	5%	$^{1}/_{20}$	.05
5)	1%	$^{1}/_{100}$	.01

Level 2 (Page 108)

	Percents	Fractions	Decimals
1)	100%	**1**	**1**
2)	375%	3 ¾	**3.75**
3)	**37.5%**	⅜	**.375**
4)	1000%	**10**	**10**
5)	**87.5%**	⅞	.875

Level 3 (Page 109)

	Percents	Fractions	Decimals
1)	**2 ½% or 2.5%**	$^{25}/_{1000}$ or ¼₀	.025
2)	107%	1 $^{7}/_{100}$	**1.07**
3)	**133 ⅓%**	1 ⅓	1.3333
4)	6.25%	¹⁄₁₆	**.0625**
5)	**93.75%**	$^{15}/_{16}$	**.9375**

Genius Level (Page 109)

	Percents	Fractions	Decimals
1)	¼%	¹⁄₄₀₀	.0025
2)	$^{7}/_{10}$%	$^{7}/_{1000}$	.007
3)	¹⁄₁₀%	¹⁄₁₀₀₀	**.001**
4)	55.5%	$^{555}/_{1000}$ or $^{111}/_{200}$	.555
5)	**$396**		
	50% of $800 = $400		
	1% of $800 = $8	½% of $800 = $4	$400 - $4 = $396
6)	.025%	$^{25}/_{100,000}$ or ¼₀₀₀	**.00025**

Answers: Metric System

(Page 111)

1) **41° F**

     $5 \times 1.8 = 9$ $9 + 32 = 41°F$

2) **0° C**

     $32 - 32 = 0$ $0 \div 1.8 = 0$

3) **161.6° F**

     $72 \times 1.8 = 129.6 + 32 = 161.6$

4) **37° C**

     $98.6 - 32 = 66.6$ $66.6 \div 1.8 = 37$

5) **32° F**

     $0 \times 1.8 = 0$ $0 + 32 = 32$

(Page 113)

1) **3461 miles**

     5572 kilometers $\div 1.61 = 3461$ miles

2) **3864 kilometers**

     2400 miles $\times 1.61 = 3864$ kilometers

3) **150,000,000 kilometers**

     $93,000,000 \times 1.61 = 149,730,000$ kilometers

4) **.62 miles**

     1 kilometer $\div 1.61 = .62$

5) **25,000 miles**

     $40,250 \div 1.61 = 25,000$ miles

(Page 114)

1) **25 decimeters**

2) **100 centimeters**

3) **1000 millimeters**

4) **2 yards**

    2 x 39.4 = 78.8 inches   36 inches per yard

5) **15.76 inches**

    4 decimeters = .4 meters   .4 meters x 39.4 = 15.76 inches

6) **5 meters**

    Reverse through the changing machine   197 inches ÷ 39.4 = 5 meters

7) **2000 millimeters**

    Reverse through the changing machine
    78.8 inches ÷ 39.4 = 2 meters   2 meters = 2000 millimeters

8) **300 millimeters**

9) **.0758 meters**

    A millimeter is $\frac{1}{1000}$ of a meter

10) **50,000,000 millimeters**

    50 kilometers x 1000 = 50,000 meters   50,000 meters x 1000 = 50,000,000 millimeters

(Page 115)

1) **220 pounds**

    100 kilograms x 2.2 = 220 pounds

2) **50 kilograms**

    110 pounds ÷ 2.2 = 50 kilograms

3) **No**

    20,000 kilograms = 44,000 pounds

4) **65 kilograms**

    143 ÷ 2.2 = 65 kilograms

5) **$52,800**

    A kilograms is 2.2 times heavier than a pound. $24,000 x 2.2 = $52,800

(Page 116)

1) **3500 milliliters**

2) **8.5 liters**

3) **4.228 quarts**
4000 milliliters = 4 liters   4 liters x 1.057 = 4.228 quarts

4) **¼ liter**

5) **1 liter**
1.057 quarts ÷ 1.057 = 1 liter

6) **4.228 quarts**
4 x 1.057 = 4.228

7) **3.78 liters**
4 quarts ÷ 1.057 = 3.78

8) **$3 per quart**
A liter is larger than a quart because one liter = more than one quart
$3 per liter is equal to $3 per 1.057 quarts
$3 per quart is more expensive than $3 per 1.057 quarts

9) **$6**
4.228 quarts ÷ 1.057 = 4 liters   4 liters x $1.50 = $6

10) **$47.30**
50 quarts ÷ 1.057 = 47.3 liters   47.3 liters x $1 = $47.30

Problem Set 1 (Page 118)

Warmup: No, water freezes at 0°C   Yes, water boils at 100°C

Level 1: 37° Celsius

  98.6 - 32 = 66.6   66.6 ÷ 1.8 = 37

Level 2: 10 degrees

  Convert 68° F to Celsius: 68 - 32 = 36   36 ÷ 1.8 = 20
  Convert 50° F to Celsius: 50 - 32 = 18   18 ÷ 1.8 = 10
  20° C - 10° C = 10 degree drop

Level 3: -273.15° Celsius

  Convert -459.67 to Celsius: -459.67 - 32 = -491.67
  -491.67 ÷ 1.8 = -273.15

Genius Level: -40°

  Try numbers in the machine until they come out the same.

Problem Set 2 (Page 119)

Warmup: 16.1 kilometers

  10 x 1.61 = 16.1 kilometers

Level 1: 66 kph

  41 x 1.61 = 66.01

Level 2: No

  100 ÷ 1.61 = 62.1 mph

Level 3: 15.5 hours

  790 miles x 1.61 = 1271.9 kilometers
  1271.9 kilometers ÷ 82 kilometers per hour = 15.51 hours

Genius Level: 9:00 A.M. on September 9th

  2790 miles x 1.61 = 4491.9 kilometers   4491.9 kilometers ÷ 100 kph = 44.92 hours
  44.92 x 60 = 2695 minutes   2695 minutes is equal to 44 hours and 55 minutes
  12:05 P.M. + 55 minutes = 1:00 P.M. September 7th
  Add 44 hours   48 hours would be September 9th at 1:00 P.M. so to get 44 hours
  subtract 4 hours from 1:00 P.M. = 9:00 A.M.

Problem Set 3 (Page 120)

Warmup: 1057 quarts

   1000 x 1.057 = 1057 quarts

Level 1: b) Slightly smaller than a liter

Level 2: Station A

   Station B charges $4 per gallon so it charges $1 per quart
   Station A charges $1 for one liter x 1.057 = 1.057 quarts $1 for 1.057 quarts
   $1 for 1.057 quarts is cheaper than $1 for one quart

Level 3: $89

   80 liters x 1.057 = 84.56 quarts   84.54 ÷ 4 = 21.14 gallons
   21.14 gallons x $4.20 per gallon = $88.79

Genius Level: 2,734,910 liters

   365 x 1980 gallons = 722,700 gallons
   722,700 gallons x 4 quarts per gallon = 2,890,800 quarts
   2,890,800 quarts ÷ 1.057 = 2,734,910 liters

Problem Set 4 (Page 121)

Warmup: 6.6 pounds

   3 kilograms x 2.2 = 6.6 pounds

Level 1: Abe

   50 kilograms x 2.2 = 110 pounds

Level 2: 10,000 kilograms

   22,000 pounds ÷ 2.2 = 10,000 kilograms

Level 3: 35.2 ounces

   A kilogram is equal to 2.2 pounds, so 16 ounces x 2.2 pounds = 35.2 ounces in a kilogram

Genius Level: No

   16 ounces in a pound.   Gold at the store sells for 16 x $1500 = $24,000 per pound.
   2.2 pounds in a kilogram so gold at the store sells for 2.2 x $24,000 = $52,800 per kilogram.

   A gram is $\frac{1}{1000}$ of a kilogram, so the store's selling price per gram is
   $52,800 ÷ 1000 = $52.80

Problem Set 5 (Page 122)

Warmup: 305 feet

1 meter = 39.4 inches 93 meters x 39.4 = 3664.2 inches
3664.2 inches ÷ 12 = 305.35

Level 1: 443 meters

1454 feet x 12 = 17,448 inches   17,448 inches ÷ 39.4 = 442.8

Level 2: 829.8 meters

82,980 centimeters ÷ 100 = 829.8 meters

Level 3: ½ mile tall

829.8 meters x 39.4 = 32,694 inches = 2725 feet.
2725 feet ÷ 5280 feet in one mile = .52   .52 is very close to .5 or ½

Genius Level: a) an aspirin

1 meter = 39.4 inches
One millimeter is 1/1000 of a meter so 39.4 ÷ 1000 = .0394 inches in one millimeter
.0394 inches in one millimeter x 11.6 millimeters = .46 inches, which is about the size of an aspirin.

Level 1 (Page 123)

1) Yes

Kilometers ÷ 1.61 = miles   85 kilometers ÷ 1.61 = 52.79 miles per hour speed limit

2) 39 feet

12 meters x 39.4 = 472.8 inches   472.8 inches ÷ 12 = 39.4 feet

3) 40˚C

104˚ - 32 = 72   72 ÷ 1.8 = 40

4) a) About the same speed

Kilometers ÷ 1.61 = miles   8 kilometers ÷ 1.61 = 4.97 miles per hour

5) Boiling: 100˚C   Freezing: 0˚C

Level 2 (Page 124)

1) c) quarter

       millimeters x .0394 = inches

       24.3 x .0394 = .96 inches   A quarter's diameter is approximately one inch.

2) 10,000 square centimeters

       There are 100 centimeters in a meter.   100 cm x 100 cm = 10,000 square centimeters

3) 299,000 kilometers per second

       miles x 1.61 = kilometers   186,000 x 1.61 = 299,460 kilometers per second

4) c) 75 liters

       20 gallons = 80 quarts   80 ÷ 1.057 = 75.7 liters

5) a) 20°C

       Room temperature is approximately 70°F   70° - 32 = 38   38 ÷ 1.8 = 21

Level 3 (Page 125)

1) 3 cows

       kilograms x 2.2 = pounds   20,000 kilograms x 2.2 = 44,000 pounds

       4000 pounds too heavy so 3 cows must be removed. (4500 pounds)

2) -37.89°F

       -38.83 x 1.8 = -69.9   -69.9 + 32 = -37.89

3) 6 liters

       Minutes in a day: 24 hours x 60 minutes per hour = 1440 minutes

       8640 liters ÷ 1440 minutes = 6 liters pumped in a minute

4) 34 cents

       miles x 1.61 = kilometers   1 mile = 1.61 kilometers   55 cents ÷ 1.61 = 34.16 cents

5) d) about ½ pound

       There are 1000 grams in a kilogram, so 200 grams is .2 kilograms.

       kilograms x 2.2 = pounds   .2 x 2.2 = .44 pounds which is a little less than a half pound

Genius Level (Page 126)

1) 1000 basketballs

11 tons x 2000 pounds in each ton = 22,000 pounds

pounds ÷ 2.2 = kilograms   22,000 pounds ÷ 2.2 = 10,000 kilograms

10,000 kilograms is the weight limit

The truck is 150 kilograms too heavy or 150 x 1000 = 150,000 grams too heavy

150,000 grams ÷ 150 grams per basketball = 1000 basketballs

2) -100

Start at absolute zero expressed in Fahrenheit and round to the nearest whole number.
-460

Add the number of degrees Fahrenheit at which water boils. -460 + 212 = -248

Add the number of degrees Fahrenheit that water freezes. -248 + 32 = -216

Add the number of degrees Fahrenheit of the normal body temperature.
-216 + 98.6 = -117.4

Finally, add 17.4 to your answer. -117.4 + 17.4 = -100

3) 37 miles per hour

kilometers ÷ 1.61 = miles   A kilometer per minute is 60 kilometers per hour.

60 kilometers ÷ 1.61 = 37.26 miles per hour

4) 5 miles per second

kilometers ÷ 1.61 = miles   27,870 kilometers per hour ÷ 1.61 = 17,310 miles per hour

3600 seconds in an hour   17,310 ÷ 3600 = 4.8 miles per second

5) No, it should cost $51.45

grams x .035 = ounces   .98 grams x .035 = .0343 ounces

.0343 ounces x $1500 per ounce = $51.45

6) Super Genius d) About 600 miles

The cubic meter is 1000 millimeters on a side. Because of this there must be
1000 x 1000 x 1000 = 1,000,000,000 cubic millimeters to put in a line.

There are 1000 millimeters per meter and 1,000,000 millimeters per kilometer.
1,000,000,000 millimeters cubes ÷ 1,000,000 per kilometer = 1000 kilometers
1000 kilometers is equal to about 600 miles.

Answers: Language of Algebra

(Page 128)

1) $60n$

2) $3n$

3) $60n$

4) $12n$

5) $n + 5$

6) $4n$

7) $10n$

8) $4n$

9) $n \div 4$ To change quarts to gallons, divide by 4

10) **Sister: $3n$**   **Ferret: n - 5**

11) n - 24   Total books: n Amount read: 24   Books left: n - 24

12) $3n + 3$ $n + (n + 1) + (n + 2) = 3n + 3$

13) $100n$ 100 centimeters in each meter

14) $8n$ Pig legs: 4 x n = $4n$ Ducks: $2n$ Duck legs: 2 x $2n$ = $4n$

15) $6n$ Width: n Length: $2n$ Perimeter: $n + n + 2n + 2n = 6n$

(Page 130)

1) **$n/36$ yards**   How many yards are in 100 inches?   100 inches ÷ 36 = yards

2) **$n/24$ days**   How many days are in 48 hours?   48 hours ÷ 24 = days

3) **$n/365$ years**   How many years are in 1000 days?   1000 days ÷ 365 = years

4) **$n/8$ gallons**   How many gallons are in 16 pints?   16 pints ÷ 8 = gallons

5) **$n/3600$ hours**   How many hours are in 10,000 seconds?   10,000 seconds ÷ 3600 = hours

Problem Set 1 (Page 131)

Warmup: $n + 2$

Level 1: $4 \times n$

Level 2: $80 - n$

Level 3: Total legs: $12n + 5$

Horses: n horse legs: $4n$

pigs: $2n$ pig legs: $4 \times 2n = 8n$ dog legs: 5

Genius Level: Duck legs: $2(100-n)$

Pigs: n pig legs: $4n$ Ducks: $100 - n$ duck legs: $2(100-n)$

Problem Set 2 (Page 132)

Warmup: $n + 3$

Level 1: $5n$

Level 2: $n \div 5$

Level 3: 175n cents

Number of quarters: $7n$ Value of quarters: $25 \times 7n = 175n$

Genius Level: $100n + 20n + 5n + 13 = 125n + 13$

Number of pennies: 13   Value: 13 cents
Number of nickels: n Value: $5n$
Number of dimes: $2n$ Value $10 \times 2n = 20n$
Number of quarters: $4n$ Value: $25 \times 4n = 100n$
$125n + 13$

Problem Set 3 (Page 133)

Warmup: $n + 1$

Level 1: $n+4$ n $n+1$ $n+2$ $n+3$ $n+4$

Level 2: $n + 8$ n $n+2$ $n+4$ $n+6$ $n+8$

Level 3: $6n + 30$

n $n+2$ $n+4$ $n+6$ $n+8$ $n+10$ Add together = $6n + 30$

Genius Level: $n + 3$

n $n+1$ $n+2$ $n+3$ $n+4$ $n+5$ $n+6$
Add = $7n + 21$ $(7n + 21) \div 7 = n + 3$

Problem Set 4 (Page 133)

Warmup: 3*n* feet   *n* x 3 = 3*n*

Level 1: 12*n* inches

Level 2: 5*n* miles

Level 3: 3600*n* seconds

60 x 60 = 3600 seconds in one hour.   3600 x *n* = 3600*n* seconds are in *n* hours.

Genius Level: *n* ÷ 5280 miles

If a slug crawls 10,000 feet, it crawls 10,000 ÷ 5280 miles
If it crawls *n* feet, it crawls *n* ÷ 5280 miles

Problem Set 5 (Page 134)

Warmup: *n* + 2

Level 1: 4*n*   4 x *n* = 4*n*

Level 2: 3*n* + 5

Level 3: *n* ÷ 8

Genius Level: Perimeter: 10*n*   **Area: 4 x *n* x *n***
Width: *n*   Length: 4*n*
Perimeter: *n* + *n* + 4*n* + 4*n* = 10*n*   Area: length x width   *n* x 4*n*

Problem Set 6 (Page 134)

Warmup: *n* + 3

Level 1: *n* - 3

Level 2: 6*n* - 3

Level 3: 3*n* + 20
Bill: *n*   Dan: 3*n*   Alicia: 3*n* + 20

Genius Level: 60 years old
Grandson: *n*   Ed: 4*n*   Son: 4*n* - 28   Daughter: 4*n* - 34
n + 4*n* + 4*n* - 28 + 4*n* - 34 = 133   13*n* - 62 = 133   13*n* = 195
n = 15   Ed = 4*n* or 60 years old

Level 1 (Page 135)

1) 2*n* legs

      If there were 10 chickens, there would be 2 x 10 = 20 legs

      There are *n* chickens, so there are 2 x *n* = 2*n* legs

2) *n* - 32

      Sara's age is Mother's age - 32   Mother's age is *n*, so Sara's age is *n* - 32

3) 25*n*

      n quarters are worth 25 cents each. Total value: *n* x 25 = 25*n* cents

4) *n* + 5

      The length is 5 inches longer than the width. The width is *n*, so the length is *n* + 5.

5) 8*n* + 12

      Spider legs: 8 legs each times *n* spiders = 8*n* spider legs

      Grasshopper legs: 6 x 2 = 12   Total legs: 8*n* + 12

Level 2 (Page 136)

1) *n* + 4

      Smallest: *n*   Next: *n* + 2   Next: *n* + 4

2) 60*n* miles

      In 2 hours, the car would travel 2 x 60 = 120 miles

      In 3 hours, the car would travel 3 x 60 = 180 miles

      In *n* hours, the car would travel *n* x 60 = 60*n* miles

3) 4*n* quarts

      4 quarts in a gallon

      If there are 5 gallons, there would be 4 x 5 = 20 quarts

      If there are 6 gallons, there would be 4 x 6 = 24 quarts

      If there are *n* gallons, there would be 4 x *n* = 4*n* quarts

4) 4*n* + 83 legs

      Pig legs: 4 x *n* = 4*n* legs   Cat legs: 4 x 20 = 80 legs   Triangle's legs: 3 legs

      Total: 4*n* + 80 + 3 = 4*n* + 83

5) 100*n* centimeters

      There are 100 centimeters in each meter.

      If a snail crawled 2 meters, he would crawl 2 x 100 = 200 centimeters

      If a snail crawled 10 meters, he would crawl 10 x 100 = 1000 centimeters

      If a snail crawled *n* meters, he would crawl *n* x 100 = 100*n* centimeters

Level 3 (Page 137)

1) 5n + 10

     Smallest: n next: $n + 1$ next: $n + 2$ next: $n + 3$ next: $n + 4$

     Add all 5 number: $n + n + 1 + n + 2 + n + 3 + n + 4 = 5n + 10$

2) 1800 - n

     1800 total cows and ostriches and n cows

     By subtracting the number of cows from 1800, you will get the number of ostriches: $1800 - n$

3) 20n

     Length is 9 times the width or 9 x $n = 9n$

     Perimeter: $n + n + 9n + 9n = 20n$

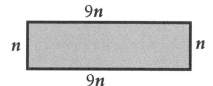

4) n ÷ 2 pairs of shoes

     If there were 10 shoes, there would be $10 ÷ 2 = 5$ pairs

     If there were 14 shoes, there would be $14 ÷ 2 = 7$ pairs

     If there were n shoes, there would be $n ÷ 2$ pairs

5)   **Shadow: n ÷ 2**   **(Shadow is half Luke's age.)**

     Luke: n

     Father: 2n   **(His father is twice Luke's age.)**

     Grandfather: n + 48   **(His grandfather is 48 years older than Luke.)**

Genius Level (Page 138)

1) 22n + 12

     Number of spiders: n Spider legs: 8 x $n = 8n$

     Number of chickens: n Chicken legs: 2 x $n = 2n$

     Number of grasshoppers: $2n$ Grasshopper legs: 6 x $2n = 12n$

     Number of cows: 3   Cow legs: 3 x 4 = 12 legs

     Total legs: $8n + 2n + 12n + 12 = 22n + 12$

2) 3600n seconds

     There are 3600 seconds in each hour. 3600 x n hours = $3600n$ seconds

3) 4(200 - n) or 800 - 4n

     Number of kangaroos: n Kangaroo legs: 2 x $n = 2n$

     Buffalo: $200 - n$ Buffalo legs: $4(200 - n)$ or $800 - 4n$

4) 12n - 4

     Perimeter: Add all 4 sides: $5n - 2 + 5n - 2 + n + n = 12n - 4$

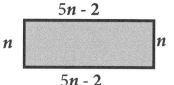

5) 21n - 5

     Grandson: n Raymond: $10n$ Son: $5n$ Daughter: $5n - 5$

     $n + 10n + 5n + 5n - 5 = 21n - 5$

Answers: Algebra Problems

(Page 143)

1) 10 snakes
Language of algebra: Snakes: n Birds: $n + 5$
Equation: $n + n + 5 = 25$ $2n + 5 = 25$ (subtract 5 from each side) $2n = 20$ $n = 10$

2) 7 years old
Language of algebra: Brian: n Brianna: $2n$
Equation: $n + 2n = 21$ $3n = 21$ $3 \times ? = 21$ $n = 7$

3) 13 dimes
Language of algebra: dimes: n Nickels: $n + 4$
Equation: $n + n + 4 = 30$ $2n + 4 = 30$ (subtract 4 from each side) $2n = 26$ $n = 13$

4) 2 years old
Language of algebra: Gabe: n Luke: $n + 19$
Equation: $n + n + 19 = 23$ $2n + 19 = 23$ (subtract 19 from each side) $2n = 4$ $n = 2$

5) 40 years old
Language of algebra: William Junior: n William: $2n$ Grandpa: $4n$
Equation: $n + 2n + 4n = 140$ $7n = 140$ $7 \times ? = 140$ $n = 20$
William = $2n$ or $2 \times 20 = 40$

(Page 143)

6) 7 cats
Language of algebra: Cats: n Ducks: $2n$ Cows: 8   Equation: $n + 2n + 8 = 29$
$3n + 8 = 29$ (Subtract 8 from both sides.) $3n = 21$ $n = 7$

7) 6 cows
Language of algebra   Cows: n Chickens: $3n$
Cow legs: $4 \times n = 4n$ Chicken legs: $2 \times 3n = 6n$ Dog legs: 6
Equation: $4n + 6n + 6 = 66$
$10n + 6 = 66$ (Subtract 6 from both sides.) $10n = 60$ $n = 6$

8) 21 years old
Language of algebra: Dwight's age: n Hiram's age: $4n$
Equation: $n + 4n = 105$ $5n = 105$ $n = 21$

9) 18 years old
Language of algebra   Rachel: n Dan: $2n$ Bria: $2n - 3$
Equation: Add ages $5n - 3 = 42$ (Add 3 to each side.)
$5n = 45$ $n = 9$ Dan is $2n$ so Dan is 18

10) 100
Language of algebra: Smallest number: n Next number: $n + 2$
Next number: $n + 4$ Next number: $n + 6$ Next number: $n + 8$
Equation: $5n + 20 = 520$ (Subtract 20 from both sides.) $5n = 500$ $n = 100$

Problem Set 1 (Page 144)

Warmup: 6 pigs

Language of algebra: Pigs: n Pig legs: $4n$ Tripod legs: 3

Equation: $4n + 3 = 27$ $4n = 24$ $n = 6$.

Level 1: 45 horses

Language of algebra: Horses: n People: $2n$

Equation: $n + 2n = 135$ $3n = 135$ $n = 45$

Level 2: 3 goats

Language of algebra: Goats: n Ducks: $7n$

Goat legs: $4 \times n = 4n$ Duck legs: $2 \times 7n = 14n$

Equation: $4n + 14n = 54$ $18n = 54$ $n = 3$

Level 3: 72 pigs

Language of algebra: Chickens: n Turkeys: $2n$ Pigs: $4n$

Equation: $n + 2n + 4n = 126$ $7n = 126$ $n = 18$

Pigs = $4n$ $4 \times 18 = 72$

Genius Level: 15 ducks

Language of algebra: Ducks: n Horses: $3n$ Dogs: 1

Duck legs: $2n$ Horse legs: $3n \times 4 = 12n$ Triangle legs: 3

Equation: $2n + 12n + 3 = 213$ $14n + 3 = 213$

(Subtract 3 from each side) $14n = 210$ $n = 15$

Problem Set 2 (Page 145)

Warmup: 4 nickels

Language of algebra:

Number of nickels: n Number of quarters: $3n$ Equation: $n + 3n = 16$

$4n = 16$ $n = 4$

Level 1: 11 pennies

Language of algebra: Pennies: n Nickels: $2n$ Dimes: $4n$

Equation: $n + 2n + 4n = 77$ $7n = 77$ $n = 11$

Level 2: 7 dimes

Language of algebra: Small pile: n dimes   Large pile: $2n$ dimes

Value of small pile in cents: $10 \times n = 10n$ Value of large pile in cents: $10 \times 2n = 20n$

Equation: $10n + 20n = 210$ $30n = 210$ $n = 7$

Level 3: 5 nickels

Language of algebra: Nickels: n Quarters: $3n$

Value of the nickels in cents: $5 \times n = 5n$ Value of the quarters in cents: $25 \times 3n = 75n$

Equation: $5n + 75n = 400$ cents $80n = 400$ $n = 5$

Genius Level: 30 quarters

Language of algebra: Nickels: n Dimes: $2n$ Quarters: $2n$ Half dollars: n

Value of nickels: $5 \times n = 5n$ Value of dimes: $2n \times 10 = 20n$

Value of quarters: $2n \times 25 = 50n$ Value of half dollars: $n \times 50 = 50n$

Equation: $5n + 20n + 50n + 50n = 1875$ $125n = 1875$ $n = 15$

Number of quarters is equal to $2n$ $2 \times 15 = 30$

Problem Set 3 (Page 146)

Warmup: 10 years old

Language of algebra: Dan: n Dave: $n + 1$

Equation: $n + n + 1 = 21$ $2n + 1 = 21$ (Subtract 1 from both sides) $2n = 20$ $n = 10$

Level 1: 8 years old

Language of algebra: Mitt: n Mack: $2n$

Equation: $n + 2n = 24$ $3n = 24$ $n = 8$

Level 2: 11 years old

Language of algebra: Stacy: n Nicki: n Lindsay: n

Equation: $n + n + n + 25 = 58$ $3n + 25 = 58$

(Subtract 25 from each side) $3n = 33$ $n = 11$

Level 3: 28 years old

Language of algebra: Holly: n Heather: $3n$ Ryan: $3n + 4$

Equation: $n + 3n + 3n + 4 = 60$ $7n + 4 = 60$ $7n = 56$

$n = 8$

Ryan's age is $3n + 4$ $3 \times 8 + 4 = 28$

Genius Level: 128 years old

Language of algebra:

Tortoise: $8n$ Bear: n Parrot: $n + 17$ Rhino: $n + 12$

Equation: Add all ages $11n + 29 = 205$

(Subtract 29 from each side) $11n = 176$ $n = 16$ Tortoise is $8n$ or $8 \times 16 = 128$

Problem Set 4 (Page 147)

Warmup: 54

Language of algebra: Smallest: n Next number: $n + 1$

Equation: $2n + 1 = 109$ $2n = 108$ $n = 54$

Level 1: 50

Language of algebra: Smallest: n Next: $n + 1$ Next: $n + 2$

Equation: $3n + 3 = 153$ $3n = 150$ $n = 50$

Level 2: 50, 51, 52, 53, 54

Language of algebra:

Smallest: n Next: $n + 1$ Next: $n + 2$ Next: $n + 3$ Largest: $n + 4$

Equation: add all 5 numbers: $5n + 10 = 260$ $5n = 250$ $n = 50$

Level 3: 41

Language of algebra: Smallest: n Next: $n + 2$ Next: $n + 4$ Largest: $n + 6$

Equation: Add all 4 numbers $4n + 12 = 176$ $4n = 164$ $n = 41$

Genius Level: 560

Language of algebra: Smallest: n Next: $n + 2$ Next: $n + 4$ Largest: $n + 6$

Equation: Add all four numbers and then divide by 4 to find the average

$(4n + 12) \div 4 = 563$ $(4n + 12) \div 4$ is the same as $n + 3$ $n + 3 = 563$ $n = 560$

Problem Set 5 (Page 148)

Warmup: 16 inches

Language of algebra:
Side: n

Equation: $4n = 64$ $n = 16$

Level 1: 18 inches

Language of algebra:

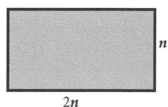

Equation: $6n = 108$ inches $n = 18$

Level 2: 10 inches

Language of algebra:

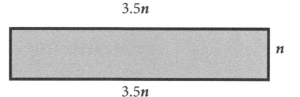

Equation: Add all sides $3.5n + 3.5n + n + n = 90$ $9n = 90$ $n = 10$

Level 3: 12 inches

Language of algebra:

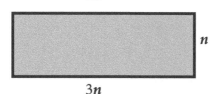

Equation: The area of a rectangle is width x length $n \times 3n = 432$
$3n^2 = 432$ Divide both sides by 3 $n^2 = 144$ $n = 12$

Genius Level: 6 inches

Language of algebra:
Side of smaller square: n Side of larger square: $3n$
Small square perimeter: $4 \times n = 4n$ Large square perimeter: $4 \times 3n = 12n$

Equation: Difference between perimeters is 48 inches
$12n - 4n = 48$ $8n = 48$ $n = 6$

Level 1 (Page 149)

1) 2.5 pounds
Language of algebra: Ferret: n Dog: $3n$
Equation: $n + 3n = 10$ $4n = 10$ $n = 2.5$

2) 60 years
Language of algebra: Bullfrog: n Catfish: $2n$
Equation: $n + 2n = 90$ $3n = 90$ $n = 30$

3) 9 quarters
Language of algebra: Number of quarters: n Number of nickels: $4n$
Equation: $5n = 45$ $n = 9$

4) 105 and 106
Language of algebra: Smallest number: n Largest number: $n + 1$
Equation: $2n + 1 = 211$ $2n = 210$ $n = 105$

5) 50 cents
Language of algebra: Bookmark: n Book: $n + 4$
Equation: $2n + 4 = 5$ $2n = 1$ dollar $n = $ half dollar or 50 cents

Level 2 (Page 150)

1) 6 inches
Language of algebra:

Equation (Add all sides): $15n = 90$ $n = 6$

2) 35 hours
Language of algebra: Hours worked: n Money earned: $12 \times n + \$75$
Equation: $12n + 75 = 495$ (Subtract 75 from each side) $12n = 420$ $n = 35$

3) 100 pounds
Language of algebra: Chimp: n Polar bear: $11n$ Killer whale: $88n$
Equation: $88n + 11n + n = 10,000$ $100n = 10,000$ $n = 100$

4) 3 years old
Language of algebra: Ole: n Shadow: $2n + 5$
Equation: $n + 2n + 5 = 14$ $3n + 5 = 14$ $3n = 9$ $n = 3$

5) 15 million
Language of algebra: Moscow: n Mexico City: $n + 5$
Equation: $2n + 5 = 35$ $2n = 30$ $n = 15$

Level 3 (Page 151)

1) 24 pounds

   Language of algebra: Weight of each brick: n Equation: $3n + 12 = 4n - 12$

   (Subtract $3n$ from each side) $12 = n - 12$ (Add 12 to both sides) $24 = n$

2) 130 pounds

   Language of algebra: Dog: n Bill: $2n + 40$ Equation: $3n + 40 = 175$

   (Subtract 40 from each side) $3n = 135$ $n = 45$ Bill: $2n + 40$ $2 \times 45 + 40 = 130$

3) $900

   Language of algebra: Mike: n Kathleen: $n + 350$ Jared: $4n$

   Equation: Add all three $6n + 350 = 5750$

   (Subtract 350 from each side) $6n = 5400$ $n = 900$

4) 20 years

   Language of algebra: Mountain lion: n Swan: $5n$ Horse: $2.5n$

   Equation: Add all three $8.5n = 170$ $n = 20$

5) 2 pounds

   Language of algebra: Polar bear: n Bison: $25n$ Elephant: $5 \times 25n = 125n$

   Equation: Add all three $151n = 302$ $n = 2$

Genius Level (Page 152)

1) $495

   Language of algebra: iPad: n Sara's money: $n - 90$ Rachel's money: $n - 405$

   Equation: Sara's money + Rachel's money = iPad cost

                $(n - 90) + (n - 405) = n$ $2n - 495 = n$

        (Subtract n from both sides and add 495 to both sides) $n = 495$

2) 48 years old

   Language of algebra: Daniel: n Luke: $3n$ Rachel: $2n$

   Equation: $6n = 96$ $n = 16$ Luke: $3n$ or $3 \times 16 = 48$

3) 15 quarters

   Language of algebra:

   Pennies: n Nickels: n Dimes: $6n$ Quarters: $3n$ Dollar bills: n

   Value of pennies: $n \times 1 = n$ Value of nickels: $n \times 5 = 5n$

   Value of dimes: $6n \times 10 = 60n$ Value of quarters: $3n \times 25 = 75n$

   Value of dollar bills: $n \times 100 = 100n$

   Equation: Add all values $241n = 1205$ cents $n = 5$ Quarters: $3n$ or $3 \times 5 = 15$

4) 60 chickens

   Language of algebra: Horses: n Chickens: $4n$ Pentagon: 1

   Horse legs: $4 \times n = 4n$ Chicken legs: $2 \times 4n = 8n$ Pentagon legs: 5

   Equation: Add all legs $4n + 8n + 5 = 185$ $12n + 5 = 185$

   (Subtract 5 from both sides) $12n = 180$ $n = 15$ Chickens: $4n$ $4 \times 15 = 60$

5) 10 million

   Language of algebra:

   Paris: n New York: $2n$ Los Angeles: $1.5n$ Tokyo: $3n$

   Equation: $7.5n = 75$ $n = 10$

Answers: Probability

(Page 157)

1) $\frac{1}{32}$

$\frac{1}{2} \times \frac{1}{2} \times \frac{1}{2} \times \frac{1}{2} \times \frac{1}{2} = \frac{1}{32}$

2) P (H,H,H,H,H)

3) $\frac{51}{52}$

51 of the 52 cards are losers.

4) $\frac{1}{36}$

First roll: $\frac{1}{6}$ Second roll: $\frac{1}{6}$ $\frac{1}{6} \times \frac{1}{6} = \frac{1}{36}$

5) $\frac{1}{6}$

First roll can be anything
Second roll: There is a $\frac{1}{6}$ probability that the second roll will match the first roll.

6) 6

1, 2, 3, 4, 5, or 6

7) $\frac{1}{6}$

There are 36 possibilities when you roll 2 dice. 6 of those add up to seven:
$1 + 6$ $6 + 1$ $5 + 2$ $2 + 5$ $3 + 4$ $4 + 3$

8) **25 families**

4 possible outcomes for having 2 children: B,B   G,G   B,G   G,B $\frac{1}{4}$ are two boys

9) **10 families**

8 possible outcomes for 3 children:
B,B,B   G,G,G   B,B,G   B,G,G   B,G,B   G,G,B   G,B,B   G,B,G $\frac{1}{8}$ have all 3 girls

10) $\frac{1}{4950}$

The first pick has a $\frac{2}{100}$ or $\frac{1}{50}$ probability of being green. If the first is green, then the next pick has a 1 in 99 probability of being green. $\frac{1}{50} \times \frac{1}{99} = \frac{1}{4950}$

Problem Set 1 (Page 158)

Warmup: ¼

When two coins are flipped, there are four possible outcomes:

H,H   H,T   T,H   T,T   Only one of the four is (heads, heads)

Level 1: ¾

When two coins are flipped, there are four possible outcomes:

H,H   H,T   T,H   T,T

Three of the four are NOT two tails

Level 2: 8

H,H,H	H,H,T	H,T,T	H,T,H
T,T,T	T,H,H	T,T,H	T,H,T

Level 3: 128 outcomes possible

$2 \times 2 \times 2 \times 2 \times 2 \times 2 \times 2 = 128$

Genius Level: $\frac{1}{1024}$

Each flip there is a ½ probability of getting heads

10 flips: $\frac{1}{2} \times \frac{1}{2} \times \frac{1}{2} \times \frac{1}{2} \times \frac{1}{2} \times \frac{1}{2} \times \frac{1}{2} \times \frac{1}{2} \times \frac{1}{2} \times \frac{1}{2} = \frac{1}{1024}$

Problem Set 2 (Page 159)

Warmup: $\frac{7}{100}$

100 marbles and only 7 are not black marbles

Level 1: a) ½

There will always be a ½ chance that a flipped coin will be tails no matter how many times it is flipped.

Level 2:

white marble, black marble   black marble, black marble   white marble, white marble

Level 3: $\frac{1}{16}$

Probability first pick is white: $\frac{50}{100}$ or ½   Probability 2nd pick is white: ½

Probability 3rd pick is white: ½   Probability 4th pick is white: ½

All 4 picks white: $\frac{1}{2} \times \frac{1}{2} \times \frac{1}{2} \times \frac{1}{2} = \frac{1}{16}$

Genius Level: $\frac{1}{210}$

To pick no black marbles, a white marble must be picked each time.

Probability 1st pick white: $\frac{4}{10}$ or $\frac{2}{5}$

Probability 2nd pick white: $\frac{3}{9}$ or $\frac{1}{3}$ (3 white marbles left and a total of 9 left)

Probability 3rd pick white: $\frac{2}{8}$ or $\frac{1}{4}$ (2 white marbles left and a total of 8 left)

Probability 4th pick white: $\frac{1}{7}$ (1 white marble left and a total of 7 left)

All picks white: $\frac{2}{5} \times \frac{1}{3} \times \frac{1}{4} \times \frac{1}{7} = \frac{2}{420} = \frac{1}{210}$

Problem Set 3 (Page 160)

Warmup: Equally likely

Each has a ⅙ probability

Level 1: ½

There are 6 possible numbers and the 1, 3, 5 are odd.   3 of the 6 are odd ³⁄₆ = ½

Level 2: ⅙

The first die can be anything. The second roll has to match the first, which is a 1 in 6 probability.

Level 3: 6

| 1 + 6 | 6 + 1 | 5 + 2 | 2 + 5 | 4 + 3 | 3 + 4 |

Genius Level: d) Jerry is six times more likely to win

The probability that Jerry will win a ferret is ⅙

The probability that Stan will win is ⅟₃₆

Stan has a ⅙ probability of rolling a 5 with his first die and a ⅙ probability of rolling a 5 with his second die. ⅙ x ⅙ = ⅟₃₆

⅙ is 6 times larger than ⅟₃₆

Problem Set 4 (Page 161)

Warmup: ¼

The dog area takes up ¼ of the circle's area.

Level 1: ⅝

The tiger and mouse area take up ⅝ of the area of the circle.

Level 2: Yes

The tiger area is 4 times larger than the mouse area. The mouse area prize is 4 times larger than the tiger area prize.

Level 3: worm: 1   mouse: 1   dog: 2   tiger: 4

Worm has an area ⅛ of the circle. ⅛ of 8 spins = 1

Mouse has an area ⅛ of the circle. ⅛ of 8 spins = 1

Dog has an area ¼ of the circle. ¼ of 8 spins = 2

Tiger has an area ½ of the circle. ½ of 8 spins = 4

Genius Level: tiger: 4   dog: 2   mouse: 1   worm: 1

Probability predicts that 4 spins will land in the tiger area, 2 in the dog area, one in the mouse area and one in the worm area.

Problem Set 5 (Page 162)

Warmup: ²⁄₆ or ⅓

There are 6 kittens to chose from. Of those 6, two are black kittens.

Level 1: ⅔

There are 6 kittens to chose from. Of those 6, four are not black kittens.

Level 2: ⅖

After the first pick, there are 5 kittens in the box and two are black kittens.

Level 3: ¹⁄₁₅

The first pick, there is a ⅓ probability of picking a black kitten.
The second pick: There are 5 kittens left and only one is black, so there is a ⅕ chance of picking a black kitten.
First pick ⅓ x second pick ⅕ = ¹⁄₁₅

Genius Level: ⅕

1st pick: ⁴⁄₆ or ⅔ chance of not picking a black kitten. If you did not pick a black kitten, then:
2nd pick: Now there are 5 kittens left and 2 black kittens, so you have a ⅗ chance of not picking a black kitten. If you did not pick a black kitten, then:
3rd pick: Now there are 4 kittens left and 2 are black, so you have a ²⁄₄ or ½ probability of not picking a black kitten.

1st pick ⅔ x 2nd pick ⅗ x 3rd pick ½ = ⁶⁄₃₀ = ⅕

Level 1 (Page 163)

1) 9 students

There is a ¹⁄₁₀ probability that the correct number is picked, so approximately ¹⁄₁₀ of the students picked the prize winning number. ¹⁄₁₀ of 90 is 9

2) ⅚

⅙ chance of winning   1 - ⅙ = ⅚ chance of losing

3) ⁹⁹⁹,⁹⁹⁹⁄₁,₀₀₀,₀₀₀

1 - ¹⁄₁,₀₀₀,₀₀₀ = ⁹⁹⁹,⁹⁹⁹⁄₁,₀₀₀,₀₀₀

4) 20 cases

1 in 3 or ⅓ of class is 10 students so ⅔ of the students would be 20 students.

5) ¹⁄₂₀

5 green marbles and a total of 100 marbles   ⁵⁄₁₀₀ or ¹⁄₂₀

Level 2 (Page 164)

1) ¼

13 of the 52 cards are spades $^{13}/_{52} = ¼$

2) $^1/_{13}$

4 aces and 52 cards $^4/_{52} = ^1/_{13}$

3) $^1/_{52}$

One ace of hearts and 52 cards

4) Flipping coins

Rolling die: $^1/_6$ chance of winning

Flipping two heads: ¼ chance of winning because there are 4 possibilities and only one is (heads, heads)   HH TT HT TH

5) $^1/_{32}$

5 flips and ½ probability that each flip will be heads: ½ x ½ x ½ x ½ x ½ = $^1/_{32}$

Level 3 (Page 165)

1) 3 ways

(6 and 4)   (4 and 6)   (5 and 5)

2) $^{25}/_{36}$

$^5/_6$ probability of not being "6" for each roll $^5/_6$ x $^5/_6$ = $^{25}/_{36}$

3) $^1/_{34}$

Number of half dollars: 10
Number of quarters: 20
Number of dimes: 50
Number of nickels: 100
Number of pennies: 500
Total coins: 680   Total number of quarters: 20 $^{20}/_{680}$ = $^1/_{34}$

4) $^1/_{216}$

Probability of 1st die being a "1": $^1/_6$ 2nd die: $^1/_6$ 3rd die: $^1/_6$
Probability of all three being "1"s $^1/_6$ x $^1/_6$ x $^1/_6$ = $^1/_{216}$

5) $^1/_6$

Each ball has a $^1/_6$ probability of being the last ball remaining.

Genius Level (Page 166)

1) 30 families

The possibilities when there are 2 children are:

Boy, Boy   Girl, Girl   Boy, Girl   Girl, Boy

¼ of these possibilities have both children boys. ¼ of 120 families = 30 families

2) 60 families

The possibilities when there are 2 children are:

Boy, Boy   Girl, Girl   Boy, Girl   Girl, Boy

½ of these possibilities have one boy and one girl. ½ of 120 families = 60 families

3) $\frac{5}{18}$

There are 36 possibilities when 2 dice are rolled.

Ways to roll a "6":   1-5   5-1   2-4   4-2   3-3
Ways to roll an "8":   2-6   6-2   5-3   3-5   4-4

$\frac{10}{36}$ probability of rolling a "6" or an "8" because there are 10 ways to roll a "6" or an "8"
$\frac{10}{36} = \frac{5}{18}$

4) $\frac{1}{24}$

The probability that "4" will be picked first is ¼.   Now there are 3 balls left.
The probability that "3" will be picked first is ⅓.   Now there are 2 balls left.
The probability that "2" will be picked first is ½.   Now there is 1 ball left.
The probability of an order of 4-3-2-1:   ¼ x ⅓ x ½ = $\frac{1}{24}$

5) $\frac{1}{270,725}$

Probability 1st card is an ace: $\frac{4}{52} = \frac{1}{13}$ 51 cards left and 3 aces
Probability 2nd card is an ace: $\frac{3}{51} = \frac{1}{17}$ 50 cards left and 2 aces
Probability 3rd card is an ace: $\frac{2}{50} = \frac{1}{25}$ 49 cards left and 1 ace
Probability 4th card is an ace: $\frac{1}{49}$
Probability of picking 4 aces: $\frac{1}{13}$ x $\frac{1}{17}$ x $\frac{1}{25}$ x $\frac{1}{49}$ = $\frac{1}{270,725}$

Answers: Ratios

(Page 169)

1) 4 feet tall

Post has a shadow 6 times its height   6 times what number is 24 feet?   Answer 4 feet

2) 60 feet tall

Thinking method: The person has a height that is ⅔ of his shadow. ⅔ of 90 feet = 60 feet.
Cross-multiplying: $7.5n = 450$ $450 \div 7.5 = 60$ feet

$$\frac{\text{5 feet (Height)}}{\text{7.5 feet (Shadow)}} \quad \mathbf{X} \quad \frac{n \text{ (Tree's height)}}{\text{90 feet (Tree's shadow)}}$$

3) 143 feet

$$\frac{\text{3.25 feet (Height)}}{\text{4 feet (Shadow)}} \quad \mathbf{X} \quad \frac{n \text{ (Building's height)}}{\text{176 feet (Tree's shadow)}}$$

$4n = 3.25 \times 176$ $4n = 572$ $n = 143$

4) 30 miles

4.5 is three times 1.5 inches so the real distance is 3 x 10 miles = 30 miles

5) 85 miles apart

$$\frac{\text{1.5 inches (Map)}}{\text{10 miles (Real)}} \quad \mathbf{X} \quad \frac{\text{12.75 inches(Map)}}{n \text{ miles (Real)}}$$

Cross-multiplying: $1.5n = 127.50$ $127.50 \div 1.5 = 85$ miles

6) 900 miles

$$\frac{\text{1.75 inches (Globe)}}{\text{100 miles (Real)}} \quad \mathbf{X} \quad \frac{\text{15.75 inches (Globe)}}{n \text{ miles (Real)}}$$

$1.75n = 15.75 \times 100$ $1.75n = 1575$ $1575 \div 1.75 = 900$ miles

(Page 170)

1) 8 grams

Each group weighs 9 + 1 = 10 grams   8 groups of 10 grams   Copper is 1 gram per group

2) 150 pounds

1050 pounds ÷ 7 pounds per group (5 + 1 + 1 = 7) = 150 groups
150 groups x 1 pound per group for gold = 150 pounds

3) 510 boys

There are 805 ÷ 50 = 17 groups of 50 students in the school. In each 50 there should be approximately 30 boys. 30 boys x 17 groups = 510 boys

4) 120 pounds

The weight of each of the 7 parts of nickel must be 8 pounds because
56 pounds ÷ 7 parts nickel = 8 pounds   There are 8 parts gold and each part weighs
8 pounds: 8 parts x 8 pounds = 64 pounds   56 pounds + 64 pounds = 120 pounds

Problem Set 1 (Page 171)

Warmup: 8 feet

Dell's shadow is twice his height. 2 x 4 feet = 8 feet

Level 1: 20 feet

Thinking method: At this time of day, the watchtower's height is 4 times the length of its shadow so the tree's height must be 4 times the length of its shadow. 4 x 5 = 20
Cross multiply: 12 feet (watchtower)/ 3 feet (shadow) = n feet (tree)/ 5 feet (tree's shadow)
Cross multiply: $3n = 60$ 3 x some number = 60   3 x 20 = 60

Level 2: 2 inches

Thinking method: The yardstick's shadow is ⅙ its length. ⅙ of 12 inches is 2 inches
Cross multiply: 36 inches (yardstick)/ 6 inches (shadow) = 12 inches (ruler)/n (shadow)
Cross multiply: 36 x n = 72   36 x some number = 72   36 x 2 = 72

Level 3: 4 feet tall

Thinking method: Change to inches: 10 foot tree is 10 feet x 12 inches per foot = 120 inches
The shadow is one foot and 3 inches or 15 inches   120 inches ÷ 15 inches = 8
The tree's height is 8 times its shadow   8 x the 6 inch post shadow = 48 inches = 4 feet
Cross multiply: 120 inches (tree)/ 15 inches (shadow) = n (post)/6 inches (shadow)
Cross multiply: 15 x n = 720   15 x some number = 720   15 x 48 = 720

Genius Level: 10 centimeters

Thinking method: A shadow of 3.6 inches is ¹⁄₁₀ of a 36 inch yardstick. Because there are 100 centimeters in a meter, ¹⁄₁₀ of 100 centimeters is 10 centimeters.

Problem Set 2 (Page 172)

Warmup: 2700 miles

   3 inches x 900 miles per inch = 2700 miles

Level 1: 800 miles

   There are 8 ½ inch pieces in 4 inches   8 x 100 miles = 800 miles

Level 2: 2700 miles

   Each inch is 240 miles. There are 11.25 inches or 11 ¼ inches
   11 inches x 240 = 2640 miles   ¼ inch = 240 ÷ 4 = 60 miles   2640 + 60 = 2700 miles

Level 3: 4500 miles

   How many 2 ½ inch pieces are in 10 feet 5 inches?
   10 feet 5 inches = 125 inches   125 inches ÷ 2.5 inches = 50
   Because there are 50   2 ½ inch pieces the distance between the cities must be:
   50 x 90 miles = 4500 miles

Genius Level: 12 ½ feet

   How many 125 mile pieces are in 25,000 miles?   25,000 miles ÷ 125 = 200
   200 x ¾ inch = 200 x .75 = 150 inches   150 inches ÷ 12 = 12.5 feet

Problem Set 3 (Page 172)

Warmup: 20 feet high

   There are four ½ inch parts in 2 inches so the flea can jump 4 times its height.
   Child: 5 feet x 4 = 20 feet

Level 1: 100 feet high

   The 5 inches high that the flea can jump is 20 times its height because there are twenty
   ¼'s in 5. (There are four ¼'s in each 1)   20 x 5 feet tall = 100 feet

Level 2: 136 feet

   How many times its height can this flea jump?
   There are four ¼'s in each 1 whole or 8 x 4 = thirty two ¼'s in 8 inches
   There are two ¼'s in ½ inch
   Flea can jump 32 + 2 = 34 times its height   Child: 4 feet x 34 = 136 feet

Level 3: ³⁄₁₆ inches high

   The person can jump half his height.   Half of the flea's ⅜ inch height = ½ of ⅜ = ³⁄₁₆

Genius Level: 344 feet

   A flea can jump how many times higher than its height?
   There are sixteen ¹⁄₁₆'s in each whole. 5 x 16 = 80
   ⅜ = ⁶⁄₁₆ so there are six ¹⁄₁₆ in ⅜   80 + 6 = 86
   The flea can jump 86 times its height. Child: 4 feet x 86 = 344 feet

Problem Set 4 (Page 173)

Warmup: 3 pounds

Level 1: 9 pounds

Ratio is 5 (gold) : 3 (silver)

The 15 pounds of gold is 5 (gold) x 3 so you must multiply the 3 (silver) x 3 = 9 pounds

5 (gold) : 3 (silver) = 15 (gold) : 9 (silver)

Level 2: 30 pounds

Each 8 pounds of the statue is 5 pounds of gold and 3 pounds of silver. How many 8 pound parts are in a 48 pound statue? 8 x 6 = 48 so there are six   8 pound parts

6 parts x 5 pounds of gold per part = 30 pounds of gold

Level 3: 100 pounds

There are 5 + 10 + 20 = 35 pounds in each part of the statue. How many 35 pound parts are in a 175 pound statue? 175 ÷ 35 = 5

5 x 20 pounds of nickel in each part = 100 pounds

Genius Level: 36 pounds

How many groups of 5 pounds of gold are in the statue? 22.5 ÷ 5 = 4.5

Must be 4.5 groups of silver (3 pounds each group)   4.5 x 3 = 13.5 pounds of silver

22.5 pounds of gold + 13.5 pounds of silver = 36 pounds

Problem Set 5 (Page 174)

Warmup: 50 boys

The sample of 10 students was half boys and half girls so we can predict that the 100 students are close to half boys and half girls.

Level 1: 6 boys

Laura picked ⅓ the names and found that there were 2 boys and 8 girls. The other ⅔ of the class will probably be close to the same: 2 boys and 8 girls for each ⅓ of the class. (There could be one boy, 2 boys, 3 boys, 4 boys or more, but each ⅓ of the class is likely made up of close to 2 boys. Each ⅓ of the class = 2 boys   3 x 2 = 6)

Level 2: 30 girls

We can predict that out of each group of 15 students there will be 5 girls. There are 90 ÷ 15 = 6 groups of 15 students. 6 groups x 5 girls each = 30 girls

Level 3: 360 boys

For every group of 100 students, we can predict that there will be about 36 boys and 64 girls. There are 10 groups of 100 students each in the school. 10 x 36 = 360 boys

Genius Level: 420 girls

How many groups of 96 students are in the school? 720 ÷ 96 = 7.5 groups

We can predict there will be 56 girls in each group of 96 students. 7.5 groups x 56 = 420 girls

Problem Set 6 (Page 175)

Warmup: One week

Twice the number of dogs, so it will only last half the time

Level 1: 6 weeks

A good way to solve this is to put a number of pounds into the problem to help the brain work better. Let's do 6 pounds. Each cat would then eat 2 pounds for the 2 week period so 6 pounds would last one cat 6 weeks.

Level 2: 30 shakes

Find how many shakes per 5 goldfish. 5 is ¼ of 20, so they need ¼ of 40 shakes = 10 shakes per 5 goldfish. 15 goldfish need 3 x 10 shakes = 30 shakes

Level 3: 62 ½ days

If 12 pounds lasts 60 days, then 60 ÷ 12 = 5 days per pound. If one pound will last 5 days, then ½ pound will last 2 ½ days.   60 days + 2 ½ days = 62 ½ days

Genius Level: 4 days

If food will last 6 days for 8 people, it will last one person 48 days.

1 person: 48 days   2 people: 48 ÷ 2 = 24 days

3 people: 48 ÷ 3 = 16 days   12 people: 48 ÷ 12 = 4 days

Level 1 (Page 176)

1) 55 feet

The flagpole is 5 times its shadow. Tree is 5 x 11 = 55 feet

2) 24 pounds

For every 5 pounds of copper, there are 8 pounds of nickel. Because the statue has 15 pounds of copper, there are three 5-pound parts of copper. 3 x 8 = 24 pounds

3) 12,500 miles

Each inch is equal to 500 miles   25 inches x 500 miles = 12,500 miles

4) 40 boys

There are four groups of 25 students in the school. If there are 10 boys in each group, then we can predict that there will be 4 x 10 = 40 boys in the school.

5) 120 feet

How many times its height can this flea jump?

There are 24 ½ inch parts in 12 inches so the flea can jump 24 times its height.

5 feet x 24 = 120 feet

Level 2 (Page 177)

1) 3200 pounds

> 2 pounds is equal to 16 x 2 = 32 ounces
> The bug can carry 32 times its weight.   100 x 32 = 3200 pounds

2) 24 pounds

> Every 10 pounds of the statue has 7 pounds of tin and 3 pounds of gold. There are eight
> 10-pound groups in an 80 pound statue. 8 x 3 pounds of gold per 10 pounds = 24 pounds

3) 72 feet

> The yardstick is 6 times the length of its shadow   (36 inches in a yard ÷ 6 inches = 6)
> 6 x 12 feet = 72 feet

4) 12 days

> 10 cats is 5 times as many as the 2 cats where the food will last 60 days.
> Because there are 5 times as many cats, the food would last ⅕ of the time.
> ⅕ of 60 days = 12 days
>
> Alternative method: Find how long food would last for one cat.
> If 2 boxes last 60 days for 2 cats, they will last 120 days for one cat.
> Now we can find how long 2 boxes of food will last for any number of cats:
> 1 cat: 120 days   2 cats: 120 ÷ 2 = 60 days
> 3 cats: 120 ÷ 3 = 40 days   10 cats: 120 ÷ 10 = 12 days

5) 84 feet

> How many ⅛'s are in 3 ½? Eight ⅛'s fit into each whole and four ⅛'s fit into ½. The flea can
> jump 3 x 8 = 24 + 4 = 28 times its height.   28 x 3 feet = 84 feet

Level 3 (Page 178)

1) 36 cups

> There are 2 + 6 + 9 = 17 total parts in the recipe.
> If there were 17 total cups in the recipe, then there would be 9 cups of oats.
> There are 68 cups in the recipe. 68 ÷ 17 = 4 groups of 17
> 4 groups x 9 cups oats per group = 36 cups of oats

2) 126 black marbles

> There are 2 parts green marbles and 3 parts black marbles in the jar. Because there are 84
> green marbles and 2 parts, each part is equal to 84 ÷ 2 = 42 marbles
> 3 parts black marbles: 3 parts x 42 marbles per part = 126 black marbles

3) ½ cup of oil

> There are 2 cups in a pint, 2 pints in a quart, and 4 quarts in a gallon.
> Therefore there must be 8 pints and 16 cups in a gallon.
> If the ratio was 16:1, then one cup of oil should be used.
> Because the ratio is 32:1, ½ cup should be used.

Level 3 continued (Page 178)

4) Yes

The height of the Statue of Liberty including the pedestal and foundation is 305 feet.
⅛ fits into 2 ½ 20 times because 2 ½ ÷ ⅛ = 20
The flea can jump 20 times its height.
The giraffe is 20 feet tall x 20 times its height = 400 feet

5) 20 boys

There are 5 boys for every 7 girls (5 groups of 50 boys and 7 groups of 50 girls)
For every 12 student names picked, you can predict that 5 boys and 7 girls will be picked.
48 is 4 groups of 12   5 boys x 4 groups of 12 = 20 boys

Genius Level (Page 179)

1) 112.5 miles

½ inch is equal to $\frac{8}{16}$ inch. If $\frac{8}{16}$ = 60 miles, then $\frac{1}{16}$ = 60 ÷ 8 = 7.5 miles
$\frac{15}{16}$ then must equal 15 x 7.5 miles = 112.5 miles

2) 1:64

There are 16 cups in a gallon.
After the ink is poured in, the container has ¼ cup of ink and 16 cups of water.
¼:16   Multiply by 4: ¼ to 16 is the same as 1 to 64

3) 380 girls

4 girls and 3 boys = 7 parts   665 ÷ 7 = 95 groups of 7
4 girls for every group, so 95 groups x 4 girls = 380 girls

4) 125 feet

A meter is 100 centimeters.
For every 15 centimeters of shadow, there is 25 centimeters of height.
For every 15 feet of shadow, there must be 25 feet of tree height.
75 has 5 groups of 15   5 groups x 25 feet = 125 feet

Alternative method:
100 centimeters/60 centimeters (shadow)   = n/75 feet (shadow)
Cross-multiply: 60n = 7500   60 x some number = 7500
60 x 125 = 7500

5) 52.5 pounds

For every 18 pounds of the statue, there are 7 pounds of gold, 6 pounds of silver, and 5 pounds of platinum.
How many 18 pound parts are there in a 189 pound statue?   189 ÷ 18 = 10.5
For each 18 pound part there are 5 pounds of platinum.   10.5 x 5 = 52.5 pounds

Answers: Measurement

(Page 181)

1) 65 cents

There are 2 pints in a quart and 4 quarts in a gallon so one gallon = 8 pints
$5.20 ÷ 8 = .65

2) 7.5 seconds

8 quarts come out of the hose in one minute or 60 seconds, so 60 seconds ÷ 8 = 7.5 seconds

3) 13.75 days

There are 5280 feet in a mile. There are 5280 ÷ 16 = 330   16-foot parts in a mile so it takes the snail 330 hours. 330 hours ÷ 24 = 13.75

4) 5 seconds

5280 feet in a mile, so it takes sound 5280 ÷ 1100 = 4.8 seconds to travel one mile

(Page 184)

1) 8 astronomical units

750,000,000 miles ÷ 93,000,000 miles in an astronomical unit = 8.06

2) 65,000 astronomical units

6 trillion miles in a light-year ÷ 93,000,000 miles per astronomical unit = 64,516

3) 670,000,000 miles

186,000 miles per second x 3600 seconds per hour = 669,600,000 miles

4) c) 1/372

250,000 ÷ 93,000,000 = $^{25}/_{9300}$ = $^{1}/_{372}$

Problem Set 1 (Page 185)

Warmup: One minute

The tub takes 15 seconds per foot to drain so it would take 60 seconds to drain.
60 seconds = 1 minute

Level 1: 20 seconds

3 inches in 60 seconds means one inch every 20 seconds. $60 \div 3 = 20$ seconds

Level 2: One quart

1 ½ gallons = 6 quarts drained every 60 seconds, which equals one quart every 10 seconds.

Level 3: ⅔

1 ½ inches per minute x 8 minutes = 12 inches of the tub is full.
$^{12}/_{18} = ⅔$ so   12 is ⅔ of 18

Genius Level: 40 minutes

The tub is 70 x 4 quarts in a gallon = 280 quarts. Because the open drain drains one quart per minute, the tub is filling at a rate of 2 gallons - one quart per minute. 2 gallons = 8 quarts so the tub is filling at a rate of 7 quarts per minute.
280 quart tub ÷ 7 per minute = 40 minutes

Problem Set 2 (Page 186)

Warmup: 8 eggs

2 eggs per dozen   48 cookies = 4 dozen   2 eggs x 4 dozen = 8 eggs

Level 1: 5 eggs

2 eggs for 10 cookies = 1 egg per 5 cookies   25 cookies would need $25 \div 5 = 5$ eggs

Level 2: ⅛ teaspoon

1 ¼ pound of cookies is half of 2 ½ pounds   Half of ¼ is ⅛

Level 3: 17 teaspoons

Find the number of teaspoons in a cup.   There are 8 fluid ounces in a cup.
There are 2 tablespoons per ounce so there are 2 x 8 = 16 tablespoons in a cup
There are 3 teaspoons per tablespoon so there are 3 x 16 = 48 teaspoons in a cup

Genius Level: Honey: 2 ½ tablespoons   Oil: ⅝ cup   Milk: 7 ½ ounces

The extra cup of flour is ¼ of the amount the recipe called for because the extra 1 cup is ¼ of 4 cups.

¼ of 2 tablespoons of honey = ½ tablespoon   2 + ½ = 2 ½
¼ of ½ cup of oil: Divide ½ cup into 4 pieces ⅛ + ⅛ + ⅛ + ⅛   ½ + ⅛ = ⅝
¼ of 6 fluid ounces of milk = 6 ÷ 4 = 1 ½ ounces of extra milk needed   6 + 1 ½ = 7 ½

Problem Set 3 (Page 187)

Warmup: 3 hours

There are 6 feet in 2 yards   Each 2 feet takes one hour

Level 1: 4 yards

6 inches an hour is a foot every 2 hours   There are 12 2-hour parts in 24 hours
The snail crawls 12 feet or 4 yards

Level 2: 9 days

The sloth will move 8 yards per hour because there are 4   15-minute parts in one hour.
There are 5280 feet in a mile ÷ 3 = 1760 yards in a mile
8 yards per hour   1760 ÷ 8 = 220   So there are 220 8-yard parts in 1760 yards
It takes the sloth 220 hours or 220 ÷ 24 hours per day = 9.166 days

Level 3: One mile

60 miles per hour is 60 miles every 60 minutes.
If a leopard runs 60 miles in 60 minutes, it runs a mile every minute.

Genius Level: 10 miles per hour

There are 10 6-minute pieces in 60 minutes (hour).
If Jackie runs a mile each 6 minutes, she will run 10 miles in an hour.

Problem Set 4 (Page 188)

Warmup: a) 1 unit

92,955,807 miles = 1 astronomical unit   93,000,000 miles is close to 92,955,807 miles

Level 1: 63,360 inches

5280 feet in a mile x 12 inches per foot = 63,360 inches

Level 2: 40 astronomical units

3,700,000,000 miles ÷ 92,955,807 miles per AU = 39.8 AU

Level 3: $\frac{1}{1760}$ of a mile

There are 5280 feet ÷ 3 = 1760 yards in a mile

Genius Level: d) 33,480 years

Light travels at 186,000 miles per second. There are 3600 seconds in an hour x 24 hours in a day = 86,400 seconds in a day.
86,400 seconds in a day x 365 days in a year = 31,536,000 seconds in a year
31,536,000 seconds in a year x 186,000 miles per second = 5,865,696,000,000 miles in a light-year
5,865,696,000,000 miles in a light-year ÷ 20,000 miles per hour = 293,284,800 hours to travel a light-year.   293,284,800 hours to travel a light-year ÷ 24 hours in a day = 12,220,200 days.   12,220,200 days ÷ 365 days in a year = 33,480 years

Problem Set 5 (Page 189)

Warmup: 4 cups

2 cups in a pint and 2 pints in a quart so there must be 4 cups in one quart.

Level 1: 6 teaspoons

There are 2 tablespoons in an ounce.   There are 3 teaspoons in each tablespoon. 2 x 3 = 6

Level 2: 40 tablespoons

The recipe calls for 2 ½ cups or 2 ½ x 8 = 20 fluid ounces

There are 2 tablespoons per fluid ounce so there are 2 x 20 = 40 tablespoons in 2 ½ cups

Level 3: 4 ounces of oil

There are 8 ounces in one cup.   There are 16 cups in one gallon.

There must be 8 ounces x 16 cups = 128 ounces in one gallon

128 ÷ 32 = 4 groups of 32 ounces in one gallon so 4 ounces of oil are needed

Genius Level: ¹/₁₉₂ of a quart

Find the number of teaspoons in one quart: 6 teaspoons in one fluid ounces x 8 ounces per cup = 48 teaspoons per cup

There are 2 cups in a pint and 2 pints in a quart so there must be 4 cups in a quart.

4 cups x 48 teaspoons per cup = 192 teaspoons in one quart

One teaspoon is equal to ¹/₁₉₂ of a quart.

Level 1 (Page 190)

1) 8 hours

There are 2 quarts in a half gallon.   4 cups in a quart = 8 cups in a 2 quart container.

Leaking at a rate of a cup an hour would take 8 hours to empty.

2) 9 cups

6 cups for 20 cookies means: 3 cups for each 10 cookies.

30 cookies require 9 cups of oats.

3) 4 ants

Each ant with its egg weighs 1 + 3 ½ = 4 ½ pounds   2 ants: 4 ½ + 4 ½ = 9 pounds

4 ants: 18 pounds   5 ants: 18 + 4 ½ = 22 ½ pounds (too heavy)

4) ¼

5) 15,840 inches

5280 feet per mile ÷ 4 = 1320 feet in ¼ mile

12 inches per foot x 1320 feet = 15,840 inches

Level 2 (Page 191)

1) 10 days

> 4 people x 2 ½ = 10 days

2) 3 pints

> There are 2 pints in a quart so one pint per 15 cupcakes is needed.
> 45 cupcakes: 45 ÷ 15 = 3 pints

3) 27 seconds

> How many 1 ⅓ parts are in 36 inches?
> 1 ⅓ + 1 ⅓ + 1 ⅓ = 4 inches which would take 3 seconds
> There are 9 4-inch parts in 36 inches   9 x 3 seconds = 27 seconds

4) ¹⁄₁₆

> There are 16 cups in a gallon so one cup is ¹⁄₁₆ of a gallon.

5) ⅛ teaspoon

> ¼ of ½ teaspoon is ⅛ teaspoon

Level 3 (Page 192)

1) 1 ¾ miles

> 10 minutes is ⅙ of 60 minutes
> She can walk ¼ mile every 10 minutes because there are six ¼ mile parts in
> 1 ½ miles   (¼ + ¼ + ¼ + ¼ + ¼ + ¼ = 1 ½)
> 1 ½ for 60 minutes + ¼ for 10 minutes = 1 ¾

2) ¹⁄₁,₀₀₀,₀₀₀

> There are 1000 millimeters in a meter and 1000 meters in each kilometer. Therefore there
> must be 1,000,000 millimeters in a kilometer.

3) 37.5 cents

> There are 2 cups in a pint and 2 pints in a quart and 4 quarts in a gallon. Therefore there are
> 16 cups in a gallon. $6 ÷ 16 = 37.5 cents

4) 44 days

> 10 feet x 12 hours = 120 feet per day   5,280 feet in one mile ÷ 120 feet per day = 44 days

5) 9 hours

> After 5 hours: Mouse has traveled 50 yards
> After 6 hours: Mouse has traveled 60 yards
> After 7 hours: Mouse has traveled 70 yards
> After 8 hours: Mouse has traveled to the 90 yard line and then back to the 80 yard line
> After 9 hours: Mouse has traveled to the 100 yard line and then back to the 90 yard line

Genius Level (Page 193)

1) 1 ½ teaspoon

Find out the number of ounces in a gallon: 16 cups in a gallon x 8 ounces per cup = 128 fluid ounces in a gallon.   4 ounces of "No Deer" per 128 ounces is the same as 1 ounce per 32 ounces of water.   (128 ÷ 4 = 32)

The amount of "No Deer" required is therefore always ⅟₃₂ of the water.

Find teaspoons in one cup of water: There are 6 teaspoons per fluid ounce so there are 6 x 8 = 48 teaspoons in a cup   ⅟₃₂ of 48 = 48 ÷ 32 = 1 ½ teaspoons

2) 1 ⅞ miles

We need to find out how far Kath walks each 15 minutes (60 minutes ÷ 4 = 15 minutes)

1 ½ is $\frac{3}{2}$ $\frac{3}{2}$ ÷ 4 is the same as saying ¼ of $\frac{3}{2}$ = ⅜

Kath walks ⅜ of a mile every 15 minutes.

1 ½ miles in 60 minutes + ⅜ mile in 15 minutes = 1 ⅞ miles in 75 minutes

3) No Rapunzel's hair can only hold 150 pounds

⅔ of 3600 = 2400 ounces that Rapunzel's hair can hold

2400 ounces ÷ 16 ounces in a pound = 150 pounds

4) ⅟₇₆₈

6 teaspoons in a fluid ounce x 8 ounces in a cup = 48 teaspoons in a cup

48 teaspoons in a cup x 2 cups in a pint = 96 teaspoons in a pint

96 teaspoons in a pint x 2 pints in a quart = 192 teaspoons in a quart

192 teaspoons in a quart x 4 quarts in a gallon = 768 teaspoons in a gallon

5) 32 minutes

4 quarts per gallon x 6 gallons = 24 quarts in 6 gallons

4 cups per quart x 24 quarts = 96 cups in a 6 gallon jug

Going in — 4 cups per minute   Going out — 1 cup per minute

There is a 3 cup gain per minute

96 cups need to be added ÷ 3 cups per minute = 32 minutes

Answers: Perimeter & Circumference

(Page 195)

1) $1350

Perimeter: 270 feet x $5 = $1350

2) 44 feet

Missing sides are 8 feet and 5 feet   7 + 15 + 2 + 7 + 8 + 5 = 44 feet

3) 225 feet

Both long sides are still 100 feet. The short sides are 12.5 feet each.
100 + 100 + 12.5 + 12.5 = 225 feet

4) One inch

1st fold: 1 + 1 + ½ + ½   2nd fold: ½ + ½ + ½ + ½
3rd fold: ½ + ½ + ¼ + ¼   4th fold: ¼ + ¼ + ¼ + ¼ = 1 inch

(Page 196)

1) 12.56 feet

4 feet x 3.14 = 12.56 feet

2) 628 hours

Distance around the lake: 3.14 x 100 feet = 314 feet   ½ foot per hour: 314 x 2 = 628 hours

3) 17.85 feet

Circumference of circle: 3.14 x 10 = 31.4   Length of ¼ circle: 31.4 ÷ 4 = 7.85 feet
Each of the straight sides are half of the diameter or 5 feet each: 5 + 5 + 7.85 = 17.85 feet

4) 32.97 inches

A 7 inch minute hand makes a circle with a 14 inch diameter in one hour.
The circumference of the circle is 14 x 3.14 = 43.96 inches. 45 minutes would be ¾ of the
circumference of the circle. ¾ of 43.96 = 32.97 inches

(Page 198)

1) 15

2) 15

3) 150

4) 41

5) 32 feet

Problem Set 1 (Page 199)

Warmup: 40 feet

15 + 15 + 5 + 5 = 40 feet

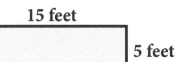

15 feet

5 feet

Level 1: 30 feet

The new rectangle is 5 feet wide and 10 feet long.

10 + 10 + 5 + 5 = 30 feet

10 feet

Level 2: 10 ½ feet

3 + 3 + 2.25 + 2.25 = 10 ½ feet

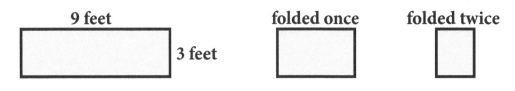

9 feet

3 feet

folded once

folded twice

Level 3: 30 inches

Each triangle is a 5 - 12 - 13 triangle

5 + 12 + 13 = 30 inches

13 inches

5 inches

12 inches

Genius Level: 17.85 inches

The circumference of the circle is 10 x 3.14 = 31.4 inches

¼ of the circumference = 31.4 ÷ 4 = 7.85 inches

Each radius = 5 inches   5 + 5 + 7.85 = 17.85 inches

5 inches

5 inches

7.85 inches

Problem Set 2 (Page 200)

Warmup: 1 hour

15 + 15 + 15 + 15 = 60 feet = 60 minutes = 1 hour

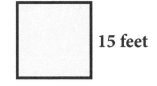

15 feet

Level 1: 1 ½ hours

The perimeter of the rectangle is 90 feet. A speed of one foot
per minute means the worm will take 90 minutes to travel
the perimeter of the rectangle. 90 minutes = 1 ½ hours

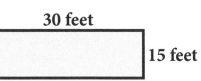

30 feet

15 feet

Problem Set 2 Continued (Page 200)

Level 2: 5 inches

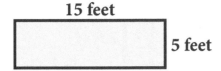

Level 3: 100 minutes

This is a 3 - 4 - 5 type of triangle so we know that the long side of the triangle is 50 inches. At the speed of ½ inch per minute, it would take 100 minutes to travel 50 inches.

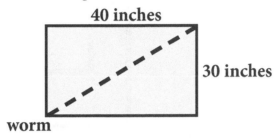

Genius Level: 40 minutes

In 40 minutes, Worm A will travel 120 inches and Worm B will travel 80 inches. The rectangle has a perimeter of 200 inches.

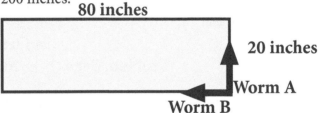

Problem Set 3 (Page 201)

Warmup: 50 feet

Missing side is 5 feet

15 + 10 + 5 + 5 + 10 + 5 = 50 feet

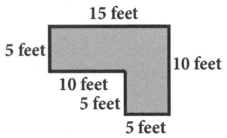

Level 1: 3 feet

15 feet + 2 feet + ?? = 20 feet

The missing side must be 3 feet

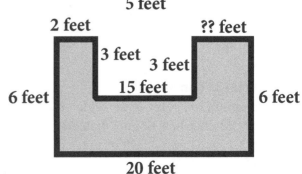

Problem Set 3 Continued (Page 201)

Level 2: 42 feet

The missing measurements are shown:
15 + 4 + 12 + 2 + 3 + 6 = 42 feet

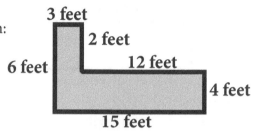

Level 3: 31 hours

62 feet around at 2 feet per hour = 31 hours

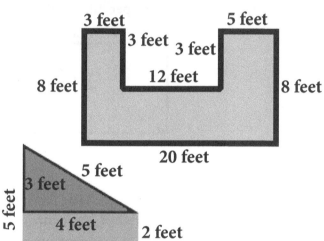

Genius Level: 16 feet

The dark shaded triangle is a 3 - 4 - 5 triangle so the missing length must be 5 feet.

5 + 5 + 4 + 2 = 16 feet

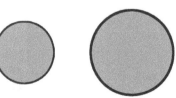

Problem Set 4 (Page 202)

Warmup: 314 inches

Circumference = pi x diameter   Circumference = 3.14 x 100 = 314 inches

Level 1: 31.4 inches

The diameter must be 10 inches   10 x 3.14 = 31.4

Level 2: pi or 3.14

The definition of pi is the ratio of circumference to diameter.

Level 3: 94.2 feet

Andy: 30 x 3.14 = 94.2 feet
Aubrey: 60 x 3.14 = 188.4 feet
188.4 - 94.2 = 94.2 feet

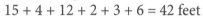

Genius Level: 12.56 inches

In an hour, the point of the 6 inch hand would travel the circumference of a 12 inch diameter circle.
12 x 3.14 = 37.68   20 minutes is ⅓ of an hour
37.68 ÷ 3 = 12.56

Problem Set 5 (Page 203)

Warmup: b) 3

Circumference of a circle is always approximately 3 times its diameter.

Level 1: 6800 miles

Circumference = Diameter x 3.14   Circumference = 2160 miles x 3.14 = 6782.4 miles

Level 2: 3.14

The definition of pi is the circumference of a circle divided by the diameter of the circle.

Level 3: 56.52 inches

In 18 hours, the hour hand travels 1 ½ times around a 12 inch diameter circle.
The circumference of a circle with a 12 inch diameter: 12 x 3.14 = 37.68

½ the circumference = 18.84 inches
37.68 + 18.84 = 56.52 inches

Genius Level: One hour

We need to find the distance across the lake. Because half of the circumference of the lake is 3.14, the distance around the lake must be 6.28 miles. To find the diameter of the circular lake:

Circumference = π x diameter   so   Diameter = Circumference ÷ π
6.28 ÷ 3.14 = 2 miles

The diameter of the lake is 2 miles. Because Jacob canoes at 2 miles per hour, it will take him one hour to canoe across the lake.

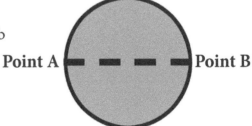

Level 1 (Page 204)

1) 10 inches

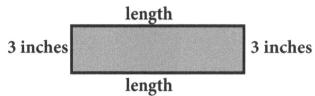

2 lengths + 3 + 3 = 26 inches
2 lengths must then equal 20 inches   Length must be 10 inches

2) 21 inches

A square has 4 equal sides.   4 x ? = 84   4 x 21 = 84

3) A = 9 inches   B = 6 inches

A is equal to 20" - 8" - 3" = 9 inches
B is equal to 10" - 4" = 6 inches

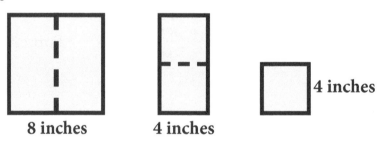

4) 16 inches

4 + 4 + 4 + 4 = 16 inches

5) 62.8 feet

Circumference = 3.14 x diameter   3.14 feet x 20 feet = 62.8 feet

Level 2 (Page 205)

1) 3 feet

> Diameter x 3.14 = Circumference   Diameter x 3.14 = 9.42 feet
> 3 feet x 3.14 = 9.42 feet (9.42 ÷ 3.14 = 3)

2) 228 feet

> Circumference of the lake: 400 feet x 3.14 = 1256 feet
> The duck needs to walk ½ the circumference: 1256 ÷ 2 = 628 feet
> 628 feet - 400 feet = 228 feet

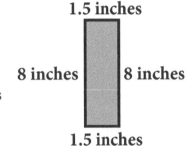

3) 19 inches

> New rectangle: 8" + 8" + 1 ½" + 1 ½" = 19 inches

4) 52 inches

> Missing sides are shown: 6" + 4" + 12" + 14" + 6" + 10" = 52 inches

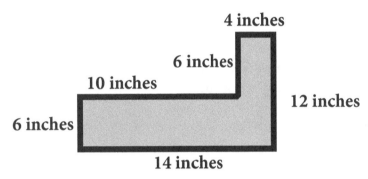

5) 4 minutes

> Circumference ÷ 3.14 = diameter   37.68 ÷ 3.14 = 12 inches
> 12 inches will take 4 minutes at a speed of 3 inches per minute.

Level 3 (Page 206)

1) 6 feet

> The circumference needs to be 3.14 x 6 people = 18.84 feet
> Diameter x pi (3.14) = 18.84 feet   18.84 ÷ 3.14 = 6 feet

2) 25.7 inches

> Circumference is pi x 10 = 31.4
> New figure's perimeter is ½ the circumference plus the diameter:
> 31.4 ÷ 2 = 15.7 + 10 = 25.7 inches

3) 1 minutes and 44 ⅔ seconds

> The circumference = 3.14 x 50 = 157 inches
> 157 inches will take 157 ÷ 1.5 = 104 ⅔ seconds
> 104 ⅔ seconds = 1 minute 44 ⅔ seconds

50 inches

4) 22 inches

> We know that the missing side is part of a 3-4-5 triangle,
> so it must be 5 inches.
> 9 + 3 + 5 + 5 = 22 inches

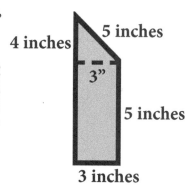

5 inches

4 inches

3"

5 inches

9 inches

3 inches

5) Swimming across the lake

> It is 400 feet across the lake. 400 feet ÷ 20 feet per minute = 20 minutes
> Circumference is 3.14 x 400 = 1256   Divide by 2 = 628 feet   628 ÷ 30 = 20.9 minutes

Genius Level (Page 207)

1) 1.9 miles

The coyote travels 5 miles in one hour and because 12 minutes is ⅕ of an hour, it must travel one mile in 12 minutes. The circumference of the lake must be 6 miles.

Diameter x 3.14 = 6 miles   Diameter = 6 miles ÷ 3.14 = 1.9 miles

2) 139.25 inches

Circumference of circle = pi x 100 inches = 314 inches
New figure is ⅛ the circumference of the circle plus 2 radiuses
314 ÷ 8 = 39.25 + 100 inches (2 radiuses) = 139.25 inches

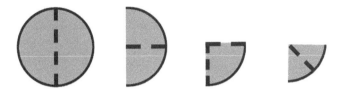

3) 5 hours and 57 minutes

If you put the semi-circles together, they make a circle with a diameter of 50 feet and a circumference of 3.14 x 50 = 157 feet
Round part of the track has a length of 157 feet
Straight part of the track has a length of 100 feet + 100 feet

The track length is therefore 157 + 100 + 100 = 357 feet
At a speed of one foot per minute, the sloth will take 357 minutes to travel the length of the track.   357 minutes ÷ 60 = 5 hours and 57 minutes

4) 314 inches or 26 feet 2 inches

The circle the hand makes has a 10 inch diameter.
Circumference = 10 inches x 3.14 = 31.4 inches traveled per hour
31.4 x 10 hours = 314 inches   314 ÷ 12 = 26 feet and 2 inches remaining

5) 3 inches

The hour hand makes 2 circles in a 24 hour period. Because it travels 37.68 inches for 2 circles, one circle must be 37.68 ÷ 2 = 18.84 inches

Circumference = Diameter x 3.14   so   Diameter = Circumference ÷ 3.14
18.84 ÷ 3.14 = 6 inches

If the diameter of the circle made by the hour hand is 6 inches, the length of the hour hand must be half of 6 inches or 3 inches.

Answers: Area

(Page 209)

1) 250 square feet

25 feet x 10 feet = 250 square feet

2) $1728

Area of room: 12 x 12 = 144 square feet
144 square feet x $12 = $1728

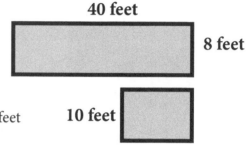

3) 470 square feet

Room splits into two rooms:
8 x 40 = 320 square feet
10 x 15 = 150 square feet   Total: 470 square feet

(Page 211)

1) 1256 square feet

Area: 3.14 x 20 feet x 20 feet = 1256 square feet

2) $125.60

$40 per 100 square feet = $40 ÷ 100 = 40 cents per square foot or $.40
Area: 3.14 x 10 x 10 = 314 square feet   314 square feet x $.40 = $125.60

3) 7536 square feet

Area of circle: 3.14 x 60 x 60 = 11,304 square feet
$120/360$ will not be painted $120/360 = 1/3$ 11,304 ÷ 3 = 3768
11,304 - 3768 = 7536 square feet will be painted

4) 21.5 square feet

The area of the square is 10 x 10 = 100 square feet
The radius of the circle = 5 feet
The area of the circle = 3.14 x 5 x 5 = 78.5 square feet
Dark green section is the area of the square (100 square feet) - the area of the circle (78.5 square feet) = 21.5 square feet

(Page 212)

1) 17.5 square inches

5" x 7" = 35   35 ÷ 2 = 17.5 square inches

2) 60 square feet

Area rectangle: 5 x 10 = 50 square feet
Height of triangle must be 4 feet
Area triangle: 5 x 4 = 20   20 ÷ 2 = 10 square feet
50 + 10 = 60 square feet

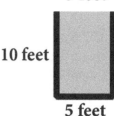

3) 6 square inches

Missing side must be 3 inches because this is a Pythagorean triple 3-4-5
3 x 4 = 12   12 ÷ 2 = 6 square inches

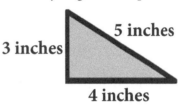

Problem Set 1 (Page 213)

Warmup: 21 Square feet

3 feet x 7 feet = 21 square feet

Level 1: 78.5 square feet

A = π x r x r   3.14 x 5 x 5 = 78.5 square feet

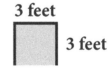

Level 2: 69 square feet

60 square feet + 9 square feet = 69 square feet

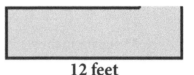

Level 3: 54 square inches

Area of the entire square: 8 inches x 8 inches = 64 square inches
Area of the missing part: 5 inches x 2 inches = 10 square inches
Area of figure: 64 - 10 = 54 square inches

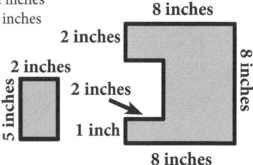

Genius Level: 2150 square feet

Area of the square: 100 feet x 100 feet = 10,000 sq. feet
Area of the circle: π x 50 x 50 = 7850 sq. feet
Area of shaded part:
10,000 square feet - 7850 square feet = 2150 sq. feet

Problem Set 2 (Page 214)

Warmup: 25 square inches

> Area of triangle is base x height ÷ 2   10 inches x 5 inches = 50 ÷ 2 = 25 square inches

Level 1: 80 square inches

> Base of 20 inches x height of 8 inches = 160   160 ÷ 2 = 80 square inches

Level 2: 127.5 square inches

> Area of the rectangle: 7 inches x 15 inches = 105 square inches
> Area of the triangle: 3 inches height x 15 inch base = 45
> 45 ÷ 2 = 22.5 square inches
> 105 + 22.5 = 127.5 square inches

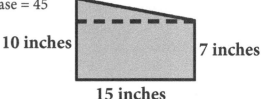

10 inches   **7 inches**

15 inches

Level 3: 72 square inches

> Area of the rectangle: 6 inches x 10 inches = 60 square inches
> Area of triangle: 6 inches x 4 inches = 24   24 ÷ 2 = 12 square inches
> Area of figure: 60 + 12 = 72 square inches

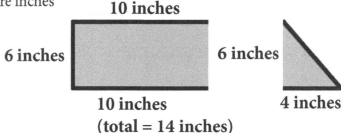

10 inches

6 inches   **6 inches**

10 inches   **4 inches**
(total = 14 inches)

Genius Level: 33 square inches

> Area of the entire triangle:
> 9 inches x 10 inches = 90 ÷ 2 = 45 square inches
> Area of rectangle:
> 3 inches x 4 inches = 12 square inches
> Area of figure:
> 45 square inches - 12 square inches = 33 square inches

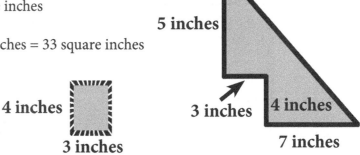

5 inches

4 inches   **3 inches**   **4 inches**

3 inches   **7 inches**

Problem Set 3 (Page 215)

Warmup: $262.50

25 square yards x $10.50 = $262.50

Level 1: $540

12 feet x 15 feet = 180 square feet   180 square feet x $3 = $540

Level 2: 432 tiles

Room is 18 feet x 24 feet = 432 square feet
Each tile has an area of one square foot.

Level 3: $1,570

The dog has ruined an area of 3.14 x 10 feet x 10 feet = 314 square feet
314 square feet x $5 = $1570

Genius Level: $320

Turn room into a rectangle and then push the 2 triangles together
Rectangle: 2 yards x 4 yards = 8 square yards
Triangle: Base of 2 yards x height of 2 yards = 4   4 ÷ 2 = 2 square yards
8 square yards + 2 square yards = 10 square yards x $32 = $320

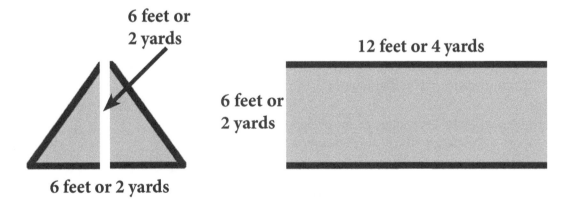

Problem Set 4 (Page 216)

Warmup: b) 2

A room 15 feet by 15 feet = 15 x 15 = 225 square feet.
Two gallons would cover 250 square feet

Level 1: 2 gallons

Area of the fence: 20 feet x 8.5 feet = 170 square feet
170 square feet ÷ 85 square feet per gallon = 2 gallons

Level 2: 3 gallons

Area of wall: 8 feet x 36 feet = 288 square feet
Area of window: 9 feet x 7 feet = 63 square feet
288 - 63 = 225 square feet of area that needs to be painted
225 ÷ 75 = 3 gallons

Level 3: 4 gallons

Area of rectangle: 12 feet x 14 feet = 168 square feet
Area of triangle: 14 feet x 8 feet = 112
112 ÷ 2 = 56 square feet
Area of wall: 168 + 56 = 224 square feet
224 square feet ÷ 56 square feet per gallon = 4 gallons

Split into a rectangle and a triangle:

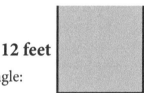

Genius Level: 47 gallons

The distance around the silo is the circumference of the circle that is the top of the silo.
Circumference = pi x diameter or 3.14 x 30 = 94.2 feet

Open silo into a rectangle as shown below:
Area = 94.2 feet x 80 feet = 7536 square feet
7536 square feet ÷ 161 square feet per gallon = 46.8 or 47 gallons

Problem Set 5 (Page 217)

Warmup: 78.5 square inches

A = π x radius x radius   Area = 3.14 x 5 x 5 = 78.5 square inches

Level 1: 19.625 square inches

A = π x radius x radius
The area of the circle is 3.14 x 5 x 5 = 78.5 square inches
90° is ¼ of the circle   78.5 square inches ÷ 4 = 19.625

Level 2: 7.85 square inches

360 degrees in a circle so 36 degrees is ¹⁄₁₀ of the circle
The area of the circle is 78.5 square inches so the missing part:
78.5 square inches ÷ 10 = 7.85 square inches

Level 3: 3.925 square inches

The area of the circle is 78.5 square inches
9° + 9° = 18°   360° ÷ 18° = 20
18 degrees is ¹⁄₂₀ of a circle   78.5 square inches ÷ 20 = 3.925 square inches

Genius Level: 29.4375 square inches

Find what fraction of the circle 135° is equal to.
135° = ¹³⁵⁄₃₆₀   135 ÷ 360 = .375
.375 x 78.5 square inches (area of circle) = 29.4375 square inches

Level 1 (Page 218)

1) $3

A rug 8 feet wide and 15 feet long is 8 x 15 = 120 square feet
$360 ÷ 120 = $3 per square foot

2) 45 square feet

One square yard is 3 feet on each side   3 x 3 = 9 square feet
5 square yards x 9 square feet per square yard = 45 square feet in 5 square yards

3) 314 square feet

Area = π x radius x radius   3.14 x 10 x 10 = 314 square feet

4) 9 tiles

Area of the tile: 4 inches x 4 inches = 16 square inches
Area of one square foot in inches: 12 inches x 12 inches = 144 square inches
144 square inches ÷ 16 square inches = 9 tiles

5) 13 feet

22 x ? = 286 square feet   286 ÷ 22 = 13

Level 2 (Page 219)

1) $2160

Change feet to yards: Room is 6 yards wide x 12 yards long

Area: 6 yards x 12 yards = 72 square yards   72 square yards x $30 = $2160

2) 200 tiles

Each tile is 1 foot x 1.5 feet = 1.5 square feet   Room is 15 feet x 20 feet = 300 square feet

300 square feet ÷ 1.5 square feet = 200 tiles

3) 300 square feet

Make 2 rectangles:

Rectangle A: 10 feet x 25 feet = 250 square feet

Rectangle B: 5 feet x 10 feet = 50 square feet

250 + 50 = 300 square feet

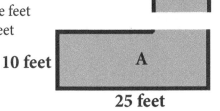

4) 48 stamps

The area of each stamp is 1.5 x 1.5 = 2.25 square inches

The area of the paper is 9 inches x 12 inches = 108 square inches

108 square inches ÷ 2.25 square inches = 48 stamps

5) $45

Room is 4 yards by 6 yards or 4 x 6 = 24 square yards $1080 ÷ 24 square yards = $45

Level 3 (Page 220)

1) 21.5 square feet

The area of the square is 10 feet x 10 feet = 100 square feet

The area of the circle is π x radius x radius or   3.14 x 5 x 5 = 78.5 square feet

The shaded part is the area of the square minus the area of the circle:

100 square feet - 78.5 square feet = 21.5 square feet

2) $2,760

Carpet cost $41.40 per square yard.

Because there are 9 square feet in each square yard, the carpet cost $41.40 ÷ 9 = $4.60 per square foot. The room is 20 feet x 30 feet = 600 square feet

600 square feet x $4.60 = $2760

3) 54 square feet of tiles

There are 6 sides of a cube. Each side is one yard x one yard which is 3 feet x 3 feet = 9 square feet.   6 sides x 9 square feet per side = 54 square feet

4) 17,280 dice

The dice each take up an area of 1 x 1 = 1 square inch. The room is 10 feet by 12 feet which is 120 inches x 144 inches   120 x 144 = 17,280 square inches

Level 3 Continued... (Page 220)

5) $70

    Area of the room wall including the window: 8 feet x 24 feet = 192 square feet

    Area of the window: 5 feet x 8 feet = 40 square feet

    Area to paint: 192 square feet - 40 square feet = 152 square feet

    152 square feet ÷ 38 square feet per quart = 4 quarts of paint needed

    4 quarts x $17.50 = $70

Genius Level (Page 221)

1) 42 square inches

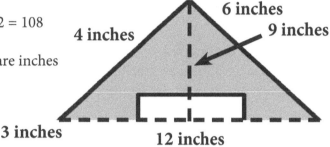

    The area of the whole triangle: 9 x 12 = 108

    108 ÷ 2 = 54 square inches

    Area of missing part: 3 x 4 = 12 square inches

    Area of figure: 54 square inches -

    12 square inches = 42 square inches

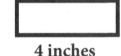

2) $432.54

    Area of each tile: 4 inch x 4 inch = 16 square inches

    Area of each side of the cube in inches: 36 inches x 36 inches = 1296 square inches

    Tiles needed for each side: 1296 square inches ÷ 16 square inches per tile = 81 tiles

    There are 6 sides to the cube: 6 x 81 tiles = 486 tiles   486 x 89 cents = $432.54

3) $1,822.50

    Area of the garden: 50 feet x 50 feet = 2500 square feet

    Area of the whole circle: π x radius x radius   3.14 x 25 x 25 = 1962.5 square feet

    Area of the semicircle: 1962.5 ÷ 2 = 981.25 square feet

    Area of unwatered part: 2500 square feet - 981.25 = 1518.75

    The cost of the rocks: 1518.75 x $1.20 per square foot = $1822.50

4) 324 square feet

    Take the rectangle out and push the triangles together.

    Now you have an area problem with a rectangle and a triangle.

    Area rectangle: 12 x 18 = 216 square feet

    Area of triangle: 18 x 12 = 216   216 ÷ 2 = 108 square feet

    Area of room: 216 + 108 = 324 square feet

5) 235.5 square feet

    Area of whole room: π x radius x radius   3.14 x 10 feet x 10 feet = 314 square feet

    Area of inside circle: 3.14 x 5 feet x 5 feet = 78.5 square feet

    Area to be painted brown: Whole room - inside circle

    314 square feet - 78.5 square feet = 235.5 square feet

Answers: Volume

(Page 223)

1) 216 cubic inches
6 inches x 6 inches x 6 inches = 216 cubic inches

2) 5,832 blocks
Volume of box is 36 inches x 36 inches x 36 inches = 46,656 cubic inches
Volume of block is 2 x 2 x 2 = 8 cubic inches   46,656 ÷ 8 = 5,832 blocks

3) 30,000 cubic decimeters
5 meters x 3 meters x 2 meters is the same as:
50 decimeters x 30 decimeters x 20 decimeters = 30,000 cubic decimeters

4) 1,000 cubes
Because there are 10 decimeters per meter, the meter box has a volume of 10 x 10 x 10 = 1000 cubic decimeters.

(Page 224)

1) One kilogram
A cubic decimeter will overflow. A cubic decimeter of water weighs one kilogram.

2) 10,000 liters
50 decimeters x 20 decimeters x 10 decimeters = 10,000 cubic decimeters
Each cubic decimeter is equal to one liter.

3) c) Cubic centimeter
A gram of water is $\frac{1}{1000}$ of a cubic decimeter of water.
A cubic decimeter weighs 1 kilogram.
A cubic decimeter is 10 centimeters on each side or 10 x 10 x 10 = 1000 cubic centimeters.

4) 1000 liters and it weighs 1000 kilograms
A cubic meter is 10 decimeters x 10 decimeters x 10 decimeters = 1000 cubic decimeters or 1000 liters and 1000 kilograms.

(Page 226)

1) 904.32 cubic inches

    3.14 x 6 inches x 6 inches x 8 inches = 904.32 cubic inches

2) 196.25

    Radius = 2.5 inches   3.14 x 2.5 inches x 2.5 inches x 10 = 196.25 cubic inches

3) 3,768 cubic inches

    Radius of hose: one inch   Length of hose: 100 feet x 12 = 1200 inches
    Volume: 3.14 x 1 inch x 1 inch x 1200 inches = 3768 cubic inches

4) 10 inches high

    Area of top x height = Volume   so   Volume ÷ area of top = Height
    785 cubic inches ÷ 78.5 square inches = 10 inches high

Problem Set 1 (Page 227)

Warmup: c) a cubic inch

Level 1: 27 cubic feet

    3 feet x 3 feet x 3 feet = 27 cubic feet

Level 2: 1728 cubic inches

    12 inches x 12 inches x 12 inches = 1728 cubic inches

Level 3: 46,656 cubic inches

    36 inches x 36 inches x 36 inches = 46,656 cubic inches

Genius Level: 147,197,952,000 blocks

    5280 feet x 5280 feet x 5280 feet = 147,197,952,000 cubic feet inside a cubic mile

Problem Set 2 (Page 228)

Warmup: 8 cubic feet

2 feet x 2 feet x 2 feet = 8 cubic feet

Level 1: 90 blocks

The volume of the box is 10 x 3 x 3 = 90 cubic feet
The volume of each block is 1 cubic foot   90 ÷ 1 = 90

Level 2: $37.50

The container is 3 yards long x 1 yard wide x 1 yard tall = 3 cubic yards
3 x $12.50 = $37.50

Level 3: 62.4 pounds per cubic foot

Change all measurements to feet: 54 inches = 4.5 feet   42 inches = 3.5 feet
8 feet x 4.5 feet x 3.5 feet = 126 cubic feet. 7862.4 ÷ 126 = 62.4 pounds for each cubic foot.

Genius Level: $2531.25

Change all measurements to yards:
81 feet = 27 yards   15 feet = 5 yards   9 inches = $\frac{9}{36}$ or ¼ of a yard

Volume: 27 x 5 x .25 = 33.75 cubic yards   33.75 x $75 = $2531.25

Problem Set 3 (Page 228)

Warmup: 1000 cubic decimeters

10 x 10 x 10 = 1000

Level 1: 1,000,000 cubic centimeters

100 centimeters are in one meter. Each side of the cube is 100 centimeters long.
100 x 100 x 100 = 1,000,000 cubic centimeters

Level 2: 30 liters

Volume of the block is 5 x 3 x 2 = 30 cubic decimeters
One cubic decimeter is equal to one liter, so 30 cubic decimeters are equal to 30 liters.

Level 3: 1000 kilograms

A cubic decimeter of water weighs one kilogram. A cubic meter's volume is 1000 cubic decimeters, so the overflowing water weighs 1000 kilograms.

Genius Level: 336 kilograms

Change measurements to decimeters.
8 x 7 x 6 = 336 cubic decimeters = 336 kilograms.

Problem Set 4 (Page 229)

Warmup: One

A liter of water is one cubic decimeter.

Level 1: 2 hours

A cubic decimeter is a liter. At a rate of a liter per minute, it will take 120 minutes to fill 120 liters. 120 minutes = 2 hours

Level 2: 1000 minutes

There are 1000 cubic decimeters in a cubic meter.
10 decimeters x 10 decimeters x 10 decimeters = 1000 cubic decimeters

Level 3: 4000 liter bottles

There are 10 x 10 x 10 = 1000 cubic decimeters in a cubic meter, so 4000 cubic decimeters go over the falls each second. Each cubic decimeter of water is a liter of water.

Genius Level: 54 pools

Each swimming pool is 500 decimeters x 250 decimeters x 20 decimeters = 2,500,000 cubic decimeters or 2,500,000 liters (because a liter of water is a cubic decimeter)

In one minute, 60 x 2,250,000 = 135,000,000 liters go over the falls
135,000,000 ÷ 2,500,000 liters per pool = 54 pools

Problem Set 5 (Page 230)

Warmup: 1000 cubic inches

The area of the top of a cylinder x height = volume
50 square inches x 20 inches = 1000 cubic inches

Level 1: 2355 cubic inches

Area of top: π x r x r = 3.14 x 5 x 5 = 78.5 square inches
78.5 square inches x 30 inches = 2355 cubic inches

Level 2: 100 inches

Area of top (12.56 square inches) x height = 1256 cubic inches
1256 ÷ 12.56 = 100   Cylinder must be 100 inches tall

Level 3: 15.7 cubic inches

The volume of the cylinder will overflow.
Area of top (π x r x r) x height = volume of cylinder 3.14 x 1 x 1 x 5 = 15.7 cubic inches

Genius Level: 942 cubic inches

Volume = π x radius x radius x height   100 feet = 1200 inches
3.14 x ½ x ½ x 1200 = 942 cubic inches

Level 1 (Page 231)

1) One foot

   1 foot x 1 foot x 1 foot = 1 cubic foot

2) 8 cubic meters

   2 meters x 2 meters x 2 meters = 8 cubic meters

3) 6 cubic yards

   6 yards x 2 yards x ½ yard = 6 cubic yards

4) 100 cubic inches

   Volume = area of top x height   10 square inches x 10 inches = 100 cubic inches

5) 10 minutes

   3 cubic feet go in each minute and 2 go out each minute, so one cubic foot of water remains each minute. It will take 10 minutes for the sink to completely fill.

Level 2 (Page 232)

1) 54 cubic feet

   Each cubic foot is 3 feet x 3 feet x 3 feet = 27 cubic feet   2 cubic yards x 27 = 54

2) $3750

   Change all measurements to yards:
   30 feet = 10 yards   45 feet = 15 yards   one foot = ⅓ yard
   10 yards x 15 yards x ⅓ yard = 50 cubic yards   50 x $75 = $3750

3) One cubic foot

   A cubic foot contains 12 inches x 12 inches x 12 inches = 1728 cubic inches

4) 10 inches

   10 inches x 10 inches x 10 inches = 1000 cubic inches

5) One hour

   180,000 cubic yards ÷ 50 cubic yards per second = 3600 seconds
   There are 3600 seconds in an hour.

Level 3 (Page 233)

1) 150 million small cubes

There are 100 centimeters in a meter, so the room is 1000 centimeters long, 500 centimeters wide and 300 meters tall.   1000 x 500 x 300 = 150,000,000 cubic centimeters

2) One cubic decimeter

A cubic decimeter is 100 millimeters long on each side.

100 millimeters x 100 millimeters x 100 millimeters = 1,000,000 cubic millimeters per cubic decimeter.

3) 9 minutes

There are 3 feet x 3 feet x 3 feet = 27 cubic feet in a cubic yard

The faucet fills one cubic foot in 20 seconds, so it will fill 3 cubic feet in 60 seconds.

It will fill 27 cubic feet in 27 ÷ 3 cubic feet per minute = 9 minutes

4) ¹⁄₁₆ of a kilogram

A cubic decimeter of water weighs one kilogram, so ¹⁄₁₆ of a cubic decimeter of water weighs ¹⁄₁₆ of a kilogram.

5) 785 cubic inches

The volume of water spilled will be equal to the volume of the cylinder.

Genius Level (Page 234)

1) 9 inches

What number multiplied by itself three time is equal to ¹⁄₆₄? ¼ x ¼ x ¼ = ¹⁄₆₄ so the length of each side of the cube is ¼ of a yard. ¼ of 36 inches is 9 inches.

2) 59 cubic inches

Volume = π x r x r x height   Diameter = ½ inch so the radius is ¼ inch

Height is 25 feet x 12 inches in a foot = 300 inches

Volume: 3.14 x ¼ x ¼ x 300 = 58.875

3) $750

7.5 feet = 2.5 yards   45 feet = 15 yards   9 inches = ¼ yard

Volume: 2.5 x 15 x .25 = 9.375 cubic yards   0.375 x $80 = $750

4) 8000 kilograms

Volume: 2 x 2 x 2 = 8 cubic meters   Water weighs one kilogram per cubic decimeter.

Each cubic meter is 10 decimeters x 10 decimeters x 10 decimeter = 1000 cubic decimeters.

8 cubic meters x 1000 = 8000 cubic decimeters

5) b) about 2 days

A cubic mile is 5280 feet x 5280 feet x 5280 feet = 147,197,952,000 cubic feet

147,197,952,000 cubic feet ÷ 850,000 cubic feet per second = 173,174 seconds

173,174 seconds ÷ 3600 seconds per hour = 48 hours or 2 days

Answers: Fun With Ratios

Level 1 (Page 236)	Level 2 (Page 237)

Level 1 (Page 236)

1) $\frac{1}{4}$

2) $\frac{1}{3}$

3) $\frac{1}{36}$

4) $\frac{1}{4}$

5) $\frac{8}{1} = 8$

6) $\frac{1}{5280}$

7) $\frac{4}{6} = \frac{2}{3}$

8) $\frac{1}{16}$

9) $\frac{1}{80}$

10) $\frac{5}{75} = \frac{1}{15}$

Level 2 (Page 237)

1) $\frac{1}{5}$
$\frac{12}{60}$ minutes $= \frac{1}{5}$

2) $\frac{1}{10}$
There are 10 decimeters in each meter

3) $\frac{1}{100}$
There are 100 millimeters in each decimeter

4) 1000
There are 1000 milliliters in each liter

5) $\frac{1}{1000}$
There are 1000 meters in each kilometer

6) $\frac{1}{9}$
A square yard is 3 feet x 3 feet = 9 square feet

7) $\frac{1}{3600}$
There are 3600 seconds in each hour

8) 12
$\frac{1}{3}$ yard = 12 inches   12 inches / 1 inch = 12

9) 1.5
$\frac{4}{2} \div \frac{8}{6} = 2 \div \frac{4}{3}$ $\frac{4}{3}$ fits into two 1.5 times

10) $\frac{1}{4}$
$\frac{4}{1} \div 16 = \frac{4}{16} = \frac{1}{4}$

Level 3 (Page 238)

1) $\frac{1}{16}$

16 cups in a gallon

2) $\frac{1}{27}$

A cubic yard is 3 feet on each side = 3 x 3 x 3 = 27 cubic feet

3) 1

Weight of a cubic decimeter of water is one kilogram

4) 1,000,000

1,000,000 millimeters in a kilometer

5) $\frac{1}{2}$

H_2O means 2 hydrogen atoms and one oxygen atom

6) 4

number of degrees of all interior angles of hexagon: 720
number of degrees of all interior angles of a triangle: 180
720 ÷ 180 = 4

7) 1

hydrogen peroxide: H_2O_2: $\frac{2}{2}$ = 1

8) 256

16 tablespoons in one cup
32 tablespoons in one pint
64 tablespoons in one quart   64 x 4 = 256 tablespoons in one gallon

9) 125

(ton/pound = 2000) ÷ (pound/ounce = 16)   2000 ÷ 16 = 125

10) 5

Height in feet of the Empire State Building (rounded to the nearest 100): 1454 or 1500
Height in feet of the Statue of Liberty (rounded to the nearest 100): 305 or 300

$\frac{1500}{300}$ = 5

Genius Level (Page 239)

1) π or 3.14

The definition of pi is the ratio of the circumference of a circle to the diameter.

2) $\frac{1}{1,000,000,000}$

Each side of a cubic meter is 1000 millimeters. 1000 x 1000 x 1000 = 1,000,000,000

3) $\frac{1}{46,656}$

A cubic yard is 36 inches on each side. 36 x 36 x 36 = 46,656

4) $\frac{1}{1.61}$

There are 1.61 kilometers in one mile.

5) 2

Octagon = 1080 total degrees   Pentagon = 540 total degrees $\frac{1080}{540} = 2$

6) $\frac{1}{27,878,400}$

A square mile is 5280 feet on each side: 5280 x 5280 = 27,878,400 square feet in one square mile

7) $\frac{1}{54}$

6 sides of a cubic yard and each side is 3 feet x 3 feet = 9 square feet
9 square feet x 6 = 54 square feet

8) 1

$\frac{1000}{8} = 125$ ÷ $\frac{2000}{16} = 125$ $125 \div 125 = 1$

9) 20

$\frac{100}{435} = \frac{20}{87}$ $\frac{20}{87}$ x 87 = 20

10) 3

$\frac{6}{18} = \frac{1}{3}$ $\frac{1}{3} \div \frac{1}{9} = 3$

Answers: Analogies
Problem Set 1 (Page 242)

Warmup: .25

25% as a decimal is .25

Level 1: .20

20% is ⅕ as fraction and .20 as a decimal

Level 2: ½₀ .05

5% is ½₀ as a fraction and .05 as a decimal

Level 3: 2.5

250% is 2.5 as a decimal

Genius Level: ¹⁄₁₀,₀₀₀ .0001

¹⁄₁₀₀% = ¹⁄₁₀₀ of 1% which is ¹⁄₁₀₀ of .01 or ¹⁄₁₀,₀₀₀   ¹⁄₁₀,₀₀₀ as a decimal is .0001

Problem Set 2 (Page 242)

Warmup: Triangle

Hexagon has 6 sides and a triangle has 3 sides

Level 1: Octagon

Level 2: Circumference

The perimeter is the outside of a rectangle. The outside of a circle is called the circumference.

Level 3: πr x r

The area of a rectangle is found by multiplying the length times the width.
The area of a circle is found by the formula πrxr.

Genius Level: Rectangular prism

Problem Set 3 (Page 243)

Warmup: One

36 inches in a yard   12 inches in a foot   One inch in an inch

Level 1: Cube

Level 2: 100

9 x 9 = 81   12 x 12 = 144   100 x 100 = 10,000

Level 3: 27

27 cubic feet in a cubic yard

Genius Level: Rhombus

A rectangle is a parallelogram with 4 corners that are each 90°.
A square is a rhombus with 4 corners that are each 90°.

Problem Set 4 (Page 243)

Warmup: 10,000
>   10 x 10 x 10 x 10 = 10,000

Level 1: 1 ¼
>   2 ½ ÷ 2 = 1 ¼

Level 2: 100,000,000 pennies
>   To find pennies, multiply dollars x 100

Level 3: 1728
>   1,728 cubic inches in a cubic foot

Genius Level: ⁴⁄₉
>   12 ÷ 3 = 4   4 ÷ 3 = 1 ⅓   1 ⅓ = ⁴⁄₃   ⁴⁄₃ ÷ 3 = ⁴⁄₉

Problem Set 5 (Page 244)

Warmup: ⁹⁸⁄₁₀₀
>   98% = ⁹⁸⁄₁₀₀

Level 1: ³⁄₂₀
>   15% = ³⁄₂₀

Level 2: πD
>   Perimeter of a rectangle is found by adding 2 lengths + 2 widths
>   Circumference of a circle is found by multiplying pi by the diameter

Level 3: 1,000
>   1,000 cubic decimeters in a cubic meter

Genius Level: 1,000,000,000
>   There are 1,000 x 1,000 x 1,000 = 1,000,000,000 cubic millimeters in a cubic meter

Problem Set 6 (Page 244)

Warmup: 75%

¾ written as a percent = 75%

Level 1: 25,000 miles

Circumference of the Earth is approximately 25,000 miles

Level 2: 360 degrees

Interior angles of a triangle = 180 degrees

Interior angles of a rectangle = 360 degrees

Level 3: 1 dollar

One trillion divided by 1,000,000 = one million   One million ÷ 1,000,000 = 1

Genius Level: $100

$100 ÷ 10,000 = 1 cent $1,000,000 ÷ 10,000 = $100

Level 1 (Page 245)

1) ½

¼ is twice ⅛   ½ is twice as large as ¼

2) 36 inches or 3 feet

3) 60

An hour is 60 times as long as a minute

4) 144

3 x 3 = 9   4 x 4 = 16   12 x 12 = 144

5) Cup

4 quarts in a gallon and 4 cups in a quart

Level 2 (Page 245)

1) ¹⁄₃₂

¹⁄₁₆ is 4 times smaller than ¼   ¹⁄₃₂ is 4 times smaller than ⅛

2) 1 minute

One hour is 12 times 5 minutes   One minute is 12 times 5 seconds

3) ½

¼ is 4 times ¹⁄₁₆   ⅛ times 4 = ½

4) kilometer

There are 5280 feet in one mile and 1000 meters in one kilometer.

5) 3

75% of $100 is ¾ of $100 or $75   300% of $100 is 3 times $100 or $300

Level 3 (Page 246)

1) 1 whole

½ is 8 times ¹⁄₁₆   ⅛ times 8 = 1

2) 1 whole

¹⁄₁₀ is 10 times as large as ¹⁄₁₀₀   1 is 10 times as large as ¹⁄₁₀

3) ½ cup

A quart is ¼ of one gallon   ½ cup is ¼ of a pint

4) 3,600

There are 3,600 seconds in one hour

5) 2%

¹⁄₅₀ is ⅕ of ¹⁄₁₀   2% is ⅕ of 10%

Genius Level (Page 246)

1) .005

100% as a decimal = 1   50% as a decimal = .5   ½% as a decimal = .005
1% as a decimal = .01   ½ of .01 = .005

2) 31,536,000

60 seconds in a minute x 60 minutes in an hour x 24 hours in a day x 365 days in a year = 31,536,000 seconds in one year

3) 1,000

2 x 2 x 2 = 8   3 x 3 x 3 = 27   10 x 10 x 10 = 1000

4) 1

1 million is ¹⁄₁,₀₀₀,₀₀₀ of a trillion   1 is ¹⁄₁,₀₀₀,₀₀₀ of a million

5) 1,000,000

A kilometer is one million millimeters.

Answers: Speed

(Page 249)

1) 7.5 hours

450 miles ÷ 60 miles per hour = 7.5 hours

2) 30 miles per hour

½ mile per 60 seconds = ½ mile per minute.

60 minutes in one hour so the car will travel: ½ mile x 60 minutes per hour = 30 miles per hour

3) 240 miles per hour

If the plane travels one mile in 15 seconds, it will travel 4 miles in one minute.

4 miles x 60 minutes in one hour = 240 miles per hour

4) 120 miles per hour

2 miles per minute x 60 minutes per hour = 120 miles per hour

5) 5 miles long

If the train took an hour to go through the crossing, it would be 20 miles long.

If it took a ½ hour to go through the crossing, it would be 10 miles long.

It took 15 minutes or ¼ hour, so the train is 20 ÷ 4 = 5 miles long.

Problem Set 1 (Page 250)

Warmup: 50 miles each hour

Level 1: 5 miles in an hour

There are five 12 minute parts to an hour: 60 ÷ 12 = 5

Level 2: One mile per hour

How many 6 minute parts are in an hour? 60 ÷ 6 = 10

10 x ¹⁄₁₀ mile = 1 whole mile

Level 3: 300 mph

One mile in 12 seconds is 5 miles each minute

60 minutes in an hour x 5 miles each minute = 300 miles per hour

Genius Level: 669,600,000 mph

3,600 seconds in one hour

3,600 x 186,000 miles per second = 669,600,000 mph

Problem Set 2 (Page 251)

Warmup: 3 yards per hour
>There are 3 yards in 9 feet

Level 1: 60 miles per hour
>60 minutes in an hour so 60 x 1 mile each minute = 60 miles per hour

Level 2: 2 inches per minute
>10 feet per hour is 10 x 12 inches in a foot = 120 inches per hour
>60 minutes in an hour so its speed is 120 ÷ 60 = 2 inches per minute

Level 3: 5 seconds
>There are 5280 feet in a mile   5280 ÷ 1100 = 4.8 seconds

Genius Level: b) 3 times as fast
>100 yards in 10 seconds = 600 yards in one minute
>There are 60 minutes in one hour so 600 yards x 60 = 36,000 yards per hour
>36,000 x 3 = 108,000 feet per hour
>108,000 feet ÷ 5280 feet per mile = 20.45 mph

Problem Set 3 (Page 252)

Warmup: 18,000 meters
>18 kilometers x 1000 meters per kilometer = 18,000 meters

Level 1: 65 kilometers
>There are 1000 meters in one kilometer so there must be 65 kilometers in 65,000 meters.

Level 2: 65,000,000 millimeters
>From looking at the chart, it is clear that there are 100 centimeters in a meter and therefore 1,000 millimeters in a meter.
>65,000 x 1000 = 65,000,000 millimeters

Level 3: 300 meters
>3 kilometers = 3000 meters in 10 hours so, it would travel 3000 meters ÷ 10 = 300 meters in an hour

Genius Level: .05 kilometers per hour or ¹⁄₂₀ of a kilometer per hour
>500 decimeters = 50 meters
>There are 1000 meters in a kilometer. What part of 1000 meters is 50 meters?
>
>50 ÷ 1000 = ¹⁄₂₀ or .05 kilometers

Problem Set 4 (Page 253)

Warmup: 3 miles per hour

The train went 3 miles in an hour so it was traveling at a speed of 3 miles per hour.

Level 1: 8 miles per hour

The train went 4 miles in half an hour so it would travel 8 miles in one hour.

Level 2: 3 miles per hour

If the train travels one mile in 20 minutes, it would travel 3 miles in 60 minutes.

Level 3: 20 miles per hour

There are 10   6-minute parts in 60 minutes so the train would travel 10 x 2 miles = 20 miles in an hour.

Genius Level: 2 minutes

The train travels 45 miles in one hour so we need to know how long it would take to travel 1 ½ miles. 45 ÷ 1 ½ = 30 so 30 trains would travel through the crossing in one hour (60 minutes).

60 minutes ÷ 30 trains = 2 minutes per train.

Problem Set 5 (Page 254)

Warmup: 10 inches per hour

20 inches in 2 hours is 10 inches every hour.

Level 1: 12 ½ inches per hour

50 inches ÷ 4 hours = 12 ½ inches per hour.

Level 2: 5 inches per minute

60 inches ÷ 12 minutes = 5 inches per minute.

Level 3: One inch per minute

The circumference is 3.14 x 50 inches = 157 inches

2 hours and 37 minutes is a total of 157 minutes

157 inches in 157 minutes is one inch per minute

Genius Level: 3 feet per hour

The diameter of the circle is 12 inches.

Circumference is 3.14 x 12 inches = 37.68 inches.

The point of the minute hand travels 37.68 inches in one hour which is about 3 feet.

Level 1 (Page 255)

1) 5 miles per hour

There are 5 12-minute parts to an hour. Each 12 minutes she travels one mile, so she would jog 1 x 5 = 5 miles in one hour

2) 3 miles per hour

There are 3 20-minute parts in an hour. 3 x one mile = 3 miles per hour

3) 1 meter per second

There are 1000 millimeters in one meter.

4) 12:00 P.M. (Midnight)

1 ½ inches per hour means 3 inches every 2 hours. Double 3 inches and there are 6 inches. This means that it will take 4 hours for 6 inches to fall.

5) 60 miles per hour

240 ÷ 4 hours = 60 miles per hour

Level 2 (Page 256)

1) 9 miles per hour

3 miles in 20 minutes would be 9 miles in 60 minutes.

2) One mile

3 minutes is ³⁄₆₀ or ¹⁄₂₀ of an hour.   ¹⁄₂₀ of 20 miles is one mile

3) 2 hours

Because Car A is traveling 10 miles per hour faster than Car B, it will gain 10 miles every hour. It will take 2 hours to gain the 20 miles.

4) 3 ½ miles

Tortoise: 1 mile per hour is ½ mile in ½ hour
Hare: 8 miles each hour is 4 miles in ½ hour
Hare has gone 4 miles minus the ½ mile that the tortoise has run so it is 3 ½ miles ahead

5) 11:30 A.M.

1 hour: 1 ¼ inches deep   2 hours: 1 ¼ + 1 ¼ = 2 ½ inches deep
If 2 hours is 2 ½ inches deep, then 4 hours would be 5 inches deep

Level 3 (Page 257)

1) 2 miles

Because the train was traveling at a speed of 20 miles per hour, a 20 mile long train would take 60 minutes (or an hour) to pass the oak tree. 6 minutes is ¹⁄₁₀ of 60 minutes so the train must be ¹⁄₁₀ of 20 miles or 2 miles long.

2) 2 miles

8 minutes is ⁸⁄₆₀ = ⁴⁄₃₀ = ²⁄₁₅ of an hour   ²⁄₁₅ of 15 miles is 2 miles

3) 64,800 miles per hour

3,600 seconds in an hour x 18 miles per second = 64,800 miles per hour

4) 180 miles per hour

There are 3600 seconds in an hour   Each 20 seconds is one mile
3600 ÷ 20 = 180 groups of 20 in 3600   180 miles in 3600 seconds

5) 3,240 miles per hour

There are 3600 seconds in an hour.
There are 3600 ÷ 5 = 720 5-second parts in 3600 seconds.
Each 5 seconds sound travels 4.5 miles: 4.5 miles x 720 = 3240 miles per hour.

Genius Level (Page 258)

1) 16 feet per second

11.2 miles per hour is 11.2 x 5280 feet = 59,136 feet per hour
There are 3600 seconds in an hour so Cernan drove the Lunar Roving Vehicle:
59,136 feet ÷ 3600 seconds in an hour = 16.43 feet per second

2) One inch per hour

The circumference of the circle the hour hand makes is diameter x pi
4 inches (diameter) x 3.14 (pi) = 12.56 inches
The tip of the hour hand takes 12 hours to travel the 12.56 inches
12.56 inches ÷ 12 hours = 1.05 inches per hour

3) 4 miles per hour

The rabbit took twice as long to go up the hill so the 3 miles per hour must be counted twice:
3 + 3 + 6 = 12   12 ÷ 3 parts = 4 miles per hour

4) 3 minutes

The tyrannosaurus gains on you 20 miles every hour. It runs 30 mph and you run only 10 mph. Gaining 20 miles every 60 minutes is the same as gaining 60 ÷ 20 = 3 minutes to gain one mile.

5) 15 miles per hour

¼ mile in 60 seconds (1 minute)   In 60 minutes she ran 60 x ¼ miles = 15 miles.

Answers: Bases

(Page 261)

2) 343   49   7   1

3) 729   81   9   1

4) 8   4   2   1

5) 512   64   8   1

(Page 262)

1) 4

    1 group of 3 + 1 group of 1 = 3 + 1 = 4

2) **201 pounds**

    Columns for base 8:   64   8   1

                                    3   1   1

    $(3 \times 64) + (1 \times 8) + (1 \times 1) = 192 + 8 + 1 = 201$ pounds

3) **$8 per week**

    Columns for base 2:   8   4   2   1

                                      1   0   0   0

    $(1 \times 8) + (0 \times 4) + (0 \times 2) + (0 \times 1) = 8$

4) **Base 7 columns:**   **49**   **7**   **1**

                                7   7

    7 groups of 7 would be 49 and 7 groups of 1 would be 7
    The number written the correct way would be 110

    1 group of 49   +   1 group of 7   +   0 groups of 1

    49   7   1
    1   1   0

(Page 263)

1) 1010

		(2)	(2)	
Base 2 columns:	16	8	4	2	1

Base 2 columns: 16   8   4   2   1
                          1   0   1   0

1 group of 8 in 10 (2 left over)
0 groups of 4 in 2 (2 left over)
1 group of 2 in 2 (0 left over)

2) 121 pounds

Columns for base 9:   81   9   1
                          1   2   1

(19) (1)

1 group of 81 in 100 (19 left over)
2 groups of 9 in 19 (1 left over)
1 group of 1 in 1 (0 left over)

3) 7 fingers

10 in base 7 is ((1 group of 7) + (0 groups of 1) = 7

4) 13 fingers

10 fingers changed to base 7:   (3)
Base 7 columns   49   7   1
                                      1   3

10 = 1 group of 7 with 3 left over
3 = 3 groups of 1

(Page 264)

2) $\frac{1}{5}$ $\frac{1}{25}$ $\frac{1}{125}$

3) $\frac{1}{6}$ $\frac{1}{36}$ $\frac{1}{216}$

4) $\frac{1}{9}$ $\frac{1}{81}$ $\frac{1}{729}$

5) $\frac{1}{3}$ $\frac{1}{9}$ $\frac{1}{27}$

Problem Set 1 (Page 265)

Warmup: 7 groups of 1000 or 7000

Level 1: 4 groups of 6 or 24

Level 2: 35

Base 6 columns: 36   6   1
                                3   5

3 groups of 6 in 23 with 5 left over   5 groups of 1 in 5

Level 3: 2,048   1,024   512   256   128   64   32   16   8   4   2   1

Genius Level: 100 in base 2 is equal to 4 in base 10

Problem Set 2 (Page 266)

Warmup: 100,000

Level 1: 25

Level 2: 36

Base 5 columns: 625   125   25   5   1
                                   1   2   1
1 group of 25   +   2 groups of 5 = 10   +   1 group of 1   25 + 10 + 1 = 36

Level 3: 344

Base 5 columns: 25   5   1
                     3   4   4

How many groups of 25 are in 99?   Answer: 3 with 24 left over
How many groups of 5 are in 24?   Answer: 4 with 4 left over
How many groups of 1 are in 4?   Answer: 4

Genius Level: 108 pounds

Base 7 columns: 49   7   1
                     2   1   3
$(2 \times 49) + (1 \times 7) + (3 \times 1) = 98 + 7 + 3 = 108$

Problem Set 3 (Page 267)

Warmup: 25

Base 5 columns:	625	125	25	5	1
			1	0	0

Level 1: 14

    2 groups of 5 + 4   10 + 4 = 14

Level 2: (4 x 25) + (3 x 5) + (2 x 1)

Base 5 columns:	625	125	25	5	1
			4	3	2

Level 3: 96

Base 5 columns:	25	5	1
	3	4	1

    3 groups of 25 = 75 + 4 groups of 5 = 20 + 1 group of 1 = 1   75 + 20 + 1 = 96

Genius Level: Fraction of base 10 person's coffee remaining: $\frac{3}{5}$
Fraction of base 5 person's coffee remaining: $\frac{1}{5}$

    .4 in base 10 means $\frac{4}{10}$ so there are $\frac{6}{10}$ remaining
    .4 in base 5 means $\frac{4}{5}$ so there is $\frac{1}{5}$ remaining

Problem Set 4 (Page 267)

Warmup: 49

Base 7 columns:	49	7	1
	1	0	0

Level 1: 41

    5 groups of 7 plus 6 = 35 + 6 = 41

Level 2: 219

Base 7 columns:	49	7	1
	4	3	2

    (4 x 49)   +   (3 x 7)   +   (2 x 1)   196 + 21 + 2 = 219

Level 3: 21

Base 2 columns:	32	16	8	4	2	1
		1	0	1	0	1

    1 group of 16 + 1 group of 4 + 1 group of 1 = 21

Genius Level: $324 + 27 + 2 + \frac{7}{9} + \frac{5}{81}$

Base 9 columns:	81	9	1	.	$\frac{1}{9}$	$\frac{1}{81}$
	4	3	2		7	5

Problem Set 5 (Page 268)

Warmup: 4

Base 4 columns: 64   16   4   1

Level 1: (2 x 16) + (0 x 4) + (2 x 1)

Base 4 columns: 16   4   1

Level 2: 30

Base 4 columns: 16   4   1
                        3   0

There are 3 groups of 4 in 12 with none left over

Level 3: $64

Base 2 columns:	64	32	16	8	4	2	1
	1	0	0	0	0	0	0

1 group of 64

Genius Level: $32,768

Base 2 columns:

32,768	16,384	8192	4096	2048	1024	512	256	128	64	32	16	8	4	2	1
1	0	0	0	0	0	0	0	0	0	0	0	0	0	0	0

Level 1 (Page 269)

1) 8000 + 600 + 70 + 8

2) 12 years old in base 7

Base 7 columns	49	7	1
		1	2

How many 7's are in 9?
1 group of 7 with 2 left over

3) 91 pounds

Base 9 columns:	729	81	9	1
		1	1	1

1 group of 81 + 1 group of 9 + 1 group of 1   81 + 9 + 1 = 91

4) 8 years old

Base 2 columns:	16	8	4	2	1
		1	0	0	0

There is 1 group of 8 in 1000 base 2, so she is 8 years old

5) 100,001 years old base 2

Base 2 columns:	128	64	32	16	8	4	2	1
		1	0	0	0	0	1

33 is 1 group of 32 with 1 left over. Because there is 1 left over, this means there are no groups of 16, 8, 4, or 2.

Level 2 (Page 270)

1) 3 groups of 7   **5 groups of 1**   **1 group of ⅐**

Base 7 columns:	49	7	1	.	⅐	¹⁄₄₉
		3	5		1	

2) 100 pounds

Base 5 columns:	25	5	1
	4	0	0

4 groups of 25 and 0 groups of 5 and 0 groups of 1 = 100

3) 121 pounds

Base 9 columns:	729	81	9	1
		1	2	1

100 contains 1 group of 81 with 19 left over.
19 has 2 groups of 9 with 1 left over which is equal to 1 group of 1.

4) 22 inches

Columns base 5:	25	5	1
		2	2

12 contains 2 groups of 5 with 2 left over   2 is equal to 2 groups of 1

5) 445 days

Columns base 9:	729	81	9	1
		4	4	5

365 is 4 groups of 81 with 41 left over.   41 is 4 groups of 9 with 5 left over. 5 is 5 groups of 1

Level 3 (Page 271)

1) ⁴⁄₁₀ + ⁵⁄₁₀₀

2) 4 feet 3 inches

4 feet in base 10 is 4 groups of 1 which is the same in base 5 - 4 groups of 1
3 inches are 3 groups of 1 in base 10 and base 5

3) 11 feet 11 inches

Base 5 columns:	25	5	1
		1	1

6 feet is 1 group of 5 and 1 left over which is 1 group of 1.
6 inches is 1 group of 5 and 1 left over which is 1 group of 1.

Level 3 Continued... (Page 271)

4) 10,111 hours and 111,000 minutes

Base 2 columns:

128	64	32	16	8	4	2	1
			1	0	1	1	1
		1	1	1	0	0	0

23 is 1 group of 16 with 7 left over   56 is 1 group of 32 with 24 left over
7 is 0 groups of 8 with 7 left over   24 is 1 group of 16 with 8 left over
7 is 1 group of 4 with 3 left over   8 is 1 group of 8 with 0 left over
3 is 1 group of 2 with 1 left over
1 is 1 group of 1

5) Base 5 with $100

Base 5 with $100: 1 group of 25   Base 2 with $10,000: 1 group of 16

Genius Level (Page 272)

1) $300 + 60 + 5 + \frac{1}{10,000}$

2) Bill has $936 more than Brianna

Bill has $1000 and Brianna has 1 group of 64 $1000 - $64 = $936

3) 11,100.1 ounces

Base 2 columns:

16	8	4	2	1	.	½	¼	⅛
1	1	1	0	0	.	1	0	0

28.5 is 1 group of 16 with 12.5 left over
12.5 is 1 group of 8 with 4.5 left over
4.5 is 1 group of 4 with .5 left over
.5 is 0 groups of 2 and 0 groups of 1 with .5 left over
.5 is ½   ½ is 1 group of ½ with 0 left over

4) $101.11

Base 2 columns:

8	4	2	1	.	½	¼	⅛
1	0	1	.	1	1	0	

5.75 is 1 group of 4 with 1.75 left over
1.75 is 0 group of 2 with 1.75 left over
1.75 is 1 group of 1 with .75 left over
.75 is ¾   ¾ is 1 group of ½ with ¼ left over
¼ is 1 group of ¼ with 0 left over

5) 6,240

Base 9 columns:

729	81	9	1
8	5	0	3

8 x 729 = 5832   5 x 81 = 405   0 x 9 = 0   3 x 1 = 3
5832 + 405 + 3 = 6240

WITHDRAWN